T0271012

Environmental Management of Air, Water, Agriculture, and Energy

Environmental Management of Air, Water, Agriculture, and Energy

Edited by
Ahmad Vasel-Be-Hagh
David S.K. Ting

CRC Press
Taylor & Francis Group
Boca Raton London New York

CRC Press is an imprint of the
Taylor & Francis Group, an **informa** business

CRC Press
Taylor & Francis Group
6000 Broken Sound Parkway NW, Suite 300
Boca Raton, FL 33487-2742

and by

CRC Press
2 Park Square, Milton Park, Abingdon, Oxon OX14 4RN

© 2020 by Taylor & Francis Group, LLC
CRC Press is an imprint of Taylor & Francis Group, an Informa business

No claim to original U.S. Government works

Printed on acid-free paper

International Standard Book Number-13: 978-0-367-18484-1 (Hardback)

Visit the Taylor & Francis Web site at
http://www.taylorandfrancis.com

and the CRC Press Web site at
http://www.crcpress.com

*To those who are making sacrifices to ensure
a bright future for the next generations*

*In memory of Zahra Naghibi, who died
in a plane crash on January 8, 2020*

Contents

Preface

As John Paul II puts it, "The earth will not continue to offer its harvest, except with faithful stewardship. We cannot say we love the land and then take steps to destroy it for use by future generations." As such, air, water, food, and energy need to be judicially managed, more so as we move forward with an expanding population and heightened living standards. Unfortunately, according to the second law of thermodynamics, we cannot not cause damage (generate entropy), as conveyed by Ting and Stagner in the opening chapter. The solution for sustaining the future is not to completely stop generating entropy, which in itself implies no life on a frozen Earth, but to choose the paths that are associated with the generation of substantially less entropy. Other elements, such as spirituality (not covered in this book) are equally essential for the well-being of the future generations. In Chapter 2, Berbel et al. emphasize that protecting the limited fresh surface water and groundwater must be part of any sustainable development goal. All levels of government must work together to improve water usage efficiency, desalination, energy efficiency, crop productivity, and water economics, management, and governance. Talking about water automatically leads one to think about feeding the new generations with water-friendly aquatic cultivation. In Chapter 3, Naghibi et al. delineate that underwater vegetation-fish farming can supply the increasing food demand while preserving the ecosystems. With the continual scarcity of land along with the shift in population into the big cities, vertical farming becomes an essential part of the food equation. In Chapter 4, Al-Kodmany points out that less traffic congestion and air pollution are some of the added benefits of incorporating vertical farming into high-rises in cities. Food prices can be further reduced because packaging and transportation would be eliminated.

In this electronic age with widespread personal cell phones and other gadgets, we can take advantage of the naturally available energy everywhere we go. In Chapter 5, Khanafer et al. detail the geometry optimization of a one-of-a-kind of piezoelectric microcantilever energy harvester. Extending intermittent renewable energy harnessing to large-scale ventures require proper storage to make them profitable. And in Chapter 6, Long and Vasel-Be-Hagh take this renewable technology one step further by incorporating the storage as part of the energy harvester. They coin the novel system as SAVER (i.e., an integrated vortex hydrokinetic energy converter).

It makes sense to deal with food, energy, and water collectively. In Chapter 7, some of the latest water-energy-food nexus lessons, experiences, and opportunities are conveyed by Mabhaudhi et al. Even though the coverage is within a southern African context, much of the know-how can be applied anywhere around the globe.

Hossain proposes a unique green technology for bettering water, energy, and the environment in Chapter 8. Capturing part of the transpired water from plants and electrolyzing it into hydrogen means green energy.

Self-sustaining urbanization as a sound environmental management measure is discussed in Chapter 9. Minaei reviews greening via improved land use; transit, including walkable neighborhoods; sustainable energy; and construction, including older infrastructures.

In the final chapter, Estévez describes biodigital innovations and genetics, calling them the fifth element. He calls everyone, architects and designers alike, to hurriedly develop sustainable and safe societies.

This generation needs to think about making a small sacrifice (i.e., conserve, reduce, and reuse), so that the future generations may savor this one and only beautiful planet Earth. This may be the time to realize what has been promulgated by good stewards such as Gaylord Nelson, "The ultimate test of man's conscience may be his willingness to sacrifice something today for future generations whose words of thanks will not be heard."

Acknowledgments

The editors are most grateful to the chapter contributors. It has been a pleasure dealing with pleasant and humble experts. The constructive comments of the anonymous reviewers have heightened the quality of the book. This book would not have become a reality without Joseph Clements and Lisa Wilford, thank you! We are absolutely indebted to Dr. Jacqueline A. Stagner, who scrutinized every word that we scrawled.

Editors

Dr. Ahmad Vasel-Be-Hagh is an Assistant Professor of Mechanical Engineering and the founder of the Fluid Mechanics Research Laboratory at Tennessee Technological University. He develops analytical models, numerical simulations, and laboratory/field experiments to add physical insight to the understanding of fluid mechanics for more accurate prediction and control of flows. Applications of his research include renewable energy generation and storage as well as fluid-structure interactions. To date, Vasel-Be-Hagh has coauthored more than twenty journal papers and book chapters. He has edited multiple books and special journal issues for major publishers, including Springer and Taylor & Francis Group.

Dr. David S.K. Ting worked on Combustion and Turbulence (Premixed Turbulent Flame Propagation) during his graduate years. He then experimented with Convection Heat Transfer and Fluid-Structure Interactions prior to joining the University of Windsor. Professor Ting is the founder of the Turbulence & Energy Laboratory and supervises students on a wide range of research projects, primarily in the energy conservation and renewable energy areas. To date, he has supervised and helped supervised more than seventy graduate students and coauthored more than 120 journal papers. Other than authoring *Basics of Engineering Turbulence* (2016) and *Engineering Combustion Essentials* (2018), Ting has coedited books with the Institute of Engineering and Technology and with Springer.

Contributors

Julio Berbel
Fac. C.C. Economicas y Empresariales
Universidad de Sevilla
Sevilla, Spain

Rupp Carriveau
Turbulence & Energy Laboratory
University of Windsor
Windsor, Ontario, Canada

Vimbayi G. P. Chimonyo
Centre for Transformative Agricultural
    and Food Systems, School of
    Agricultural, Earth & Environmental
    Sciences
University of KwaZulu-Natal
Pietermaritzburg, South Africa

Alberto T. Estévez
Universitate Internacional de Catalunya
Barcelona, Spain

Alfonso Expósito
Fac. C.C. Economicas y Empresariales
Universidad de Sevilla
Sevilla, Spain

Carlos Gutiérrez-Martín
Fac. C.C. Economicas y Empresariales
Universidad de Sevilla
Sevilla, Spain

Md. Faruque Hossain
Green Globe Technology
Flushing, New York

Khalil Khanafer
Mechanical Engineering Department
Australian College of Kuwait
Kuwait

and

Advanced Manufacturing Lab (AML),
    School of Engineering
University of Guelph
Guelph, Ontario, Canada University

Kheir Al-Kodmany
Department of Urban Planning and
    Policy
College of Urban Planning and Public
    Affairs
University of Illinois at Chicago
Chicago, Illinois

Stanley Liphadzi
Water Research Commission of South
    Africa
Pretoria, South Africa

Cody S. Long
Fluid Mechanics Research Laboratory
Mechanical Engineering Department
Tennessee Technological University
Cookeville, Tennessee

Tafadzwanashe Mabhaudhi
Centre for Transformative Agricultural
    and Food Systems
School of Agricultural, Earth &
    Environmental Sciences
University of KwaZulu-Natal
Scottsville, Pietermaritzburg, South
    Africa

Ali Al-Masri
Mechanical Engineering Department
Australian College of Kuwait
Kuwait

Negin Minaei
University of Windsor
Windsor, Ontario, Canada, Canada

Albert T. Modi
Centre for Transformative Agricultural
    and Food Systems
School of Agricultural, Earth &
    Environmental Sciences
University of KwaZulu-Natal
Pietermaritzburg, South Africa

Sylvester Mpandeli
Water Research Commission of South
    Africa
Pretoria, South Africa

and

School of Environmental Sciences
University of Venda
Thohoyandou, South Africa

Zahra Naghibi
Turbulence & Energy Laboratory
University of Windsor
Windsor, Ontario, Canada

Dhesigen Naidoo
Water Research Commission of South
    Africa
Lynnwood Manor, South Africa

Luxon Nhamo
International Water Management
    Institute, Southern Africa
    (IWMI-SA)
Silverton, Pretoria, South Africa

Carlos D. Pérez-Blanco
Fac. C.C. Economicas y Empresariales
Universidad de Sevilla
Sevilla, Spain

Aidan Senzanje
School of Engineering
University of KwaZulu-Natal
Pietermaritzburg, South Africa

Jacqueline A. Stagner
Turbulence & Energy Laboratory
University of Windsor
Windsor, Ontario, Canada

Kambiz Vafai
Mechanical Engineering Department
University of California
Riverside, California

1 Acknowledging the Essentials

David S.K. Ting and Jacqueline A. Stagner

CONTENTS

1.1 INTRODUCTION

Air, water, food, and energy are essentials for the survival of the inhabitants on the planet we call Earth. The list of elements of this book—air, water, food, and energy—is evidently not exhaustive. Spirituality is presumably also indispensable to the well-being of the human species. For example, in a critical review, Clark and Hunter (2019) concluded that there is clear evidence in the literature that shows correlations between spirituality and mental health and quality-of-life factors. Therefore, they encourage practicing nurses to attend to meaning, purpose, and connectedness for the spiritual well-being of the patients because this can have a profound impact on patients' overall wellness. To highlight but a specific case, Canada et al. (2019) affirm that spiritual well-being is not only important for people with cancer but also remains essential for most survivors of cancer. To contrast spiritual well-being against today's evil, electronic screen, Lee and Jirasek (2019) found that spiritual well-being positively enhances the lives of adolescents. On the other hand, electronic screen time has a negative impact on the spiritual well-being of adolescents. Although this essential is outside the scope of the current book, it is imperative if human life on Earth is to continue.

Returning to the somewhat more tangible essentials covered in this book (i.e., air, water, food, and energy): These four life-supporting elements are intrinsically connected with the environment. As an illustration, fruits and vegetables require water to flourish. This is also the case for plants that are used to feed livestock. Naturally,

1

plants interact directly with the environment and air. Thus far, air has been considered largely separated from water, food, and to a somewhat lesser extent, energy. This may be partly because air has a long history of being treated as an important stand-alone subject.

1.2 THE SECOND LAW OF THERMODYNAMICS

A famous saying concerning the laws of thermodynamics goes like this:

1. You cannot win, you can only break even.
2. You can only break even at absolute zero.
3. You cannot reach absolute zero.

The first statement says that no matter what we do, the best solution is "treatment" (i.e., not a "cure"). In thermodynamic language, according to the second law of thermodynamics, the best that we can do is not to generate entropy. This "generate no entropy," according to the second statement, can only occur at zero degrees Kelvin. It is a good thing that we cannot reach absolute zero, the third statement, because there would be no life on Earth, and we would not have been here to generate entropy in the first place.

Realistically, not many, if any, would want to go back to the simple living standards that we had a couple of centuries ago. Conservation is necessary, but expecting everyone to practice it is an elusive dream. Then, what is the solution, especially knowing that we are but entropy generators, continuously generating entropy that damages the planet Earth and beyond? It is true that every process produces entropy. Equally true is that some processes create substantially less entropy than others for providing the same desired output. This later truth grants hope. Every human being is granted to savor life, and possibly everyone should. To ensure future generations continue to have this privilege, we need to reduce, reuse, and recycle, whenever possible. The larger responsibility lays on the shoulders of the leaders, including engineers, architects, and policy makers. The different experts contributing to this book have conveyed many alternatives that are considerably more sustainable in water usage, food production, and energy conversion. We need to have a paradigm shift to these better courses and continue our efforts in deriving superior means.

1.3 WATER

Without water there is neither food nor life. Most interestingly, Shackleton et al. (2018) found that green infrastructure (an approach to manage water to mimic the natural water cycle) promotes both the spiritual and mental well-being of the residents. In Chapter 2, Berbel et al. focuses on fresh water from both surface and groundwater resources. They estimate a 60% increase in water withdrawal by 2050. As such, sustainable development goals must include adequate water for future generations. To do so, water-related innovations, such as water usage efficiency, desalination, energy efficiency, crop productivity, along with water economics (increase pricing), management, and governance at local, national, and international levels, should be enhanced. They acknowledge that long-term water sustainability is a serious global challenge that is extremely difficult to deal with.

Any "water solution" that excludes the ever-prevailing, petrochemical-related water usage is, at best, a temporary fix.

1.4 FOOD

Our sustenance depends on food, and food generation leans on nutrients, in addition to water and air. Sutton et al. (2013) warned that the escalating and affluent population, mingled with the increase in consumption of energy and animal products, will heighten nutrient losses, pollution levels, and land degradation. Consequently, the quality of water, air, and soils will deteriorate, adversely affecting climate and biodiversity. This large, encompassing "nutrient nexus" must be properly addressed, in which a critical objective is to protect the marine environment from land-based activities. The increase in nutrients such as nitrogen and phosphorus, which comes from mining of finite phosphate rock deposits, is expected to be most drastic in traditional and developing countries. Sutton et al. (2013) attribute this to population growth and an increase in income and subsequent rise in meat and dairy consumption. It is imperative that preventive actions be taken to mitigate the depletion of high-grade phosphate rock reserves, which carries with it soil losses and dispersion of soil-bound substances into the air and water (D'Odorico et al., 2018).

Positive measures to ensure healthy food for future generations include proper application of underwater vegetation along with aquatic cultivation, as discussed in Chapter 3 by Naghibi et al. Underwater farming is a propitious alternative to conventional land-based farming, and its advantages are multiplied when properly coupled with fish farming. This environmentally friendly vegetation-fish cycle can contribute to the growing need of food while preserving the ecosystems. It is worth noting that the produce from underwater is relatively more nutritious. This solution can also be extended to flood mitigation by making use of flood-prone areas for water-surface–based vegetation. With the omnipresence of water for the plant, the energy demand is low. Renewable energy also appears more accessible and possibly easier to harness.

Vertical farming, discussed in Chapter 4 by Al-Kodmany, is promising because it makes more efficient use of farmland. This is particularly the case in suburban, or even urban, settings (see Food-Energy-Water Nexus section). Some obvious fringe benefits of incorporating vertical farming into high-rises in cities include reduced traffic congestion, and thus, harmful pollution. Without the packaging and transportation costs, food prices are expected to be highly competitive. This is especially the case when the latest greenhouse technologies, such as hydroponics (with nutrient-film technique, wick system, water culture, ebb and flow, or drip feed), aeroponics, and aquaponics, are exploited. The discussion also underscores the use of natural rainwater and readily available solar energy. Al-Kodmany ends the chapter with prospects for further advancements, listing many of the existing frontiers, including Green Sense Farms, AeroFarms, Metropolis Farms, Plenty, VerticalHarvest, Lufa Farms, and VertiCropTM.

1.5   ENERGY

Moving forward, the number of Joules of energy required to produce a unit of food must be taken more seriously; similarly, the amount of water required in producing energy for human use needs to be reduced. As importantly, a progressively larger proportion of the required energy for sustaining comfortable living standards ought to come from renewable sources. Renewable and sustainable energy has been making significant inroads. Although the dominating focus is on large, commercial-scale energy harvesting, this grid-distributed energy may not be the best when it comes to powering the widespread and drastic upsurge in small-scale, high-tech electronic gadgets. For the small-scale needs, it makes sense to capitalize on naturally available energy within the environment that the devices are being used. Such is the case conveyed in Chapter 5 by Khanafer et al., which discusses geometry optimization of a piezoelectric microcantilever energy harvester. For larger, commercial-scale renewable energy harnessing, some effective means of energy storage are a must. This is because the available energy is intermittent, and so is the energy demand, and the two are often not in sync. Long and Vasel-Be-Hagh convey that the energy storage can be furthered by integrating the storage with the energy harvester. They highlight the added value of combining a vortex hydrokinetic energy converter with an unwater compressed air energy storage system (i.e., the integrated vortex hydrokinetic energy converter [SAVER]) in Chapter 6.

1.6   FOOD-ENERGY-WATER NEXUS

The nexus among food, energy, and water (FEW) has recently been widely and well recognized and studied. In Chapter 7, Mabhaudhi et al. give a detailed overview of FEW progress within the southern Africa context.

1.6.1   Urbanization

Assuming that agriculture is well taken care of by farmers, city slickers, nevertheless, are still faced with the FEW challenge within the urban context. From processing to distribution centers to consumption points, much energy and water are necessary. Zimmerman et al. (2018) presented some dynamic urban FEW models capable of accounting for the shift to healthier, but more water-intensive, foods and the changes in (severe) weather events. According to the United Nation (2018), 55% of the world's population currently resides in urban areas, and this is projected to increase to 68% by 2050. To highlight, there will be 416 million more urban dwellers in India in 2050 and 255 and 189 million in China and Nigeria, respectively (2018). Thus, it is imperative that all stakeholders and decision makers acknowledge these essentials and collaborate on devising proper measures to accommodate the impending transformations. In China, the drastic advancement of the food systems has resulted in an increase in food-related energy consumption by 53% per capita between 2002 and 2012 (Song et al., 2019). Within this studied time period, the energy share of farming actually decreased. Unfortunately, this savings is more than outweighed by that associated with food processing and, more so, transportation.

1.6.2 Globalization

Rightly asserted by de Amorim et al. (2018), the FEW nexus demands sustainable, integrated, and intelligent management on a global scale. These essential resources—water, energy, and food—are susceptible to risks imposed by economic bubbles, deflation, and failure; severe weather events; large human migrations along with improper planning; infectious disease outbreaks; and information-infrastructure breakdown. They call for international collaboration to implement measures to ensure these borderless, essential, life-supporting elements for the next generations.

1.6.3 Competitions among the Essentials

D'Odorico et al. (2018) stress that the availability of water is a key constraint in meeting food and energy needs of the increasing population and standard of living. Within the FEW nexus, food and energy (e.g., biofuel) contest each other for water. These competitions tend to further the problem. Is there a solution to the looming threat? Thankfully, there are a few promising measures. One of these is to shift food consumption patterns (e.g., eating less red meats and more sustainable, plant-based foods). Putting more emphasis on nutrition is the right way forward. This path can lead to better health and a more sustainable FEW future. Reducing food waste is a simple yet powerful step. Because of the synergy among the three systems, improving one of them can result in multiplied improvements across the board. Therefore, waste capture and recycling in the circular economy, among other actions, can substantially improve the resilience of FEW security at the global scale. These points are echoed throughout the chapter written by Mabhaudhi et al. They assert that the FEW nexus furnishes opportunities to concurrently realize FEW securities, among other benefits.

1.7 ADDITIONAL CONSIDERATIONS

A unique technical idea is proposed by Hossain, in Chapter 8: The idea is to capture water transpired from plants, using plastic tanks and the help of the force of static electricity, noting that the plants only use about 0.5% of the absorbed water while transpiring the rest. The water is then treated via ultraviolet technology. Part of the water can be harnessed as hydrogen molecules, via electrolysis, to be burned as green energy.

What about being self-sufficient, to relieve some of the emerging pressure from the different fronts, such as climate change, that confront us? This is the topic of Chapter 9, which conveys self-sustaining urbanization and self-sufficient cities in the era of climate change. In this chapter, Minaei reviews land-use configuration, transit-oriented development, and walkable neighborhoods, in addition to sustainable energy and construction. Also discussed are ways to regenerate old infrastructures to greener ones. Proper planning and implementation are needed to effectively realize the objective.

The book would have been incomplete without the enlightenment by Estévez on the fifth element—biodigital innovations and genetics, which are covered in Chapter 10. Undeniably, there is a certain amount of truth in the new great evil of our time, planetary unsustainability. And who would argue against Estévez that humans have the responsibility to mitigate this human problem? Among others, architects and designers are called to develop, with urgency, sustainable and safe societies. To not "spoil the movie," the "story" is left to the environmentally inclined readers to savor.

REFERENCES

Canada, A.L., Murphy, P.E., Stein, K.D., Alcaraz, K.I., Fitchett, G., "Trajectories of spiritual well-being in long-term survivors of cancer: A report from the American Cancer Society's studies of cancer survivors," *Cancer*, 125(10): 1726–1736, 2019.

Clark, C.C., Hunter, J., "Spirituality, spiritual well-being, and spiritual coping in advanced heart failure," *Journal of Holistic Nursing*, 37(1): 56–73, 2019.

de Amorim, W.S., Valduga, I.B., Ribeiro, J.M.P., Williamson, V.G., Krauser, G.E., Magtoto, M.K., de Andrade Guerra, J.B.S.O., "The nexus between water, energy, and food in the context of the global risks: An analysis of the interactions between food, water, and energy security," *Environmental Impact Assessment Review*, 72: 1–11, 2018.

D'Odorico, P., Davis, K.F., Rosa, L. et al., "The global food-energy-water nexus," *Reviews of Geophysics*, 56: 456–531, 2018.

Lee, S., Jirasek, I., "Associations between screen-based activity, spiritual well-being, and life satisfaction among adolescents," *Journal of Religion and Health*, 58(3): 795–804, 2019.

Shackleton, C.M., Blair, A., De Lacy, P., Kaoma, H., Mugwagwa, N., Dalu, M.T., Walton, W., "How important is green infrastructure in small and medium-sized towns? Lessons from South Africa," *Landscape and Urban Planning*, 180: 273–281, 2018.

Song, F., Reardon, T., Tian, X., Lin, C., "The energy implication of China's food system transformation," *Applied Energy*, 240: 617–629, 2019.

Sutton, M.A., Bleeker, A., Howard, C.M., et al., "Our Nutrient World: The challenge to produce more food and energy with less pollution," Global Overview of Nutrient Management. Centre for Ecology and Hydrology, Edinburgh on behalf of the Global Partnership on Nutrient Management and the International Nitrogen Initiative, 2013.

United Nations, Department of Economic and Social Affairs, News, May 16, 2018, New York, https://www.un.org/development/desa/en/news/population/2018-revision-of-world-urbanization-prospects.html, accessed on April 18, 2019.

Zimmerman, R., Zhu, Q., Dimitri, C., "A network framework for dynamic models of urban food, energy and water systems (FEWS)," *Environmental Progress & Sustainable Energy*, 37(1): 122–131, 2018.

2 Water, Where Do We Stand?

Julio Berbel, Alfonso Expósito,
Carlos Gutiérrez-Martín, and Carlos D. Pérez-Blanco

CONTENTS

2.1 INTRODUCTION

Water is essential for human welfare and for sustaining Earth's ecosystems. This natural resource constitutes a critical input for all economic sectors, but it is also a significant element for a wide variety of cultural and societal purposes. Population growth, changes in land-use patterns (e.g., agricultural intensification, urbanization), and globalization all incur significant human pressure on the availability of fresh water, both from surface and groundwater resources, thus influencing its qualitative and quantitative status. The world as we know it today is threatened by a growing and unbalanced demand for water, increasing water scarcity and pollution, climate change, and the effects of inadequate water management instruments and water governance decisions. Most of these threats and challenges are present on a global scale, thus requiring an international plan of action. Nevertheless, society generally faces significant difficulties when dealing with global challenges, particularly when these challenges require bilateral or multilateral cooperation between countries. This is illustrated by the slow development of international agreements for the improvement of global governance of water resources. Water has at the latest been considered a crucial resource for sustainable development since the United

Nations (UN) Commission on Sustainable Development adopted the declaration "Strategic Approaches to Freshwater Management" in 1998 (Cosgrove and Loucks 2015). In recent years, the concept of global water governance has grown in relevance in the international political agenda and has triggered a wide range of declarations and cooperation initiatives developed by various international organizations (e.g., Food and Agriculture Organization [FAO], Organisation for Economic Co-operation and Development [OECD], UN, World Bank). Several UN Sustainable Development Goals (SDGs) related to water provide good examples. These initiatives aim to include the consideration of water-resource management within the wider political objectives of specific action programs and to assure sustainable, efficient, and equitable development thereof.

Not only does this chapter offer a brief but complete overview of the current scenario of where we stand regarding the use of water resources on a global scale and the challenges to be faced in the near future, but it also discusses the governance and management solutions that can be used to help face these challenges. After this introductory section, this chapter analyzes the current status of water resources on a global scale, as well as the observed current demand (Section 2.2) and supply trends (Section 2.3). Potential future scenarios and challenges are also discussed. Future challenges affecting world water resources include the impacts of climate change and its associated risks (e.g., higher frequency of drought and flood events); water resources also occupy a significant place because before middle of the twenty-first century, global warming is likely to reach a 1.5°C increase unless decisive action is taken by governments worldwide (Intergovernmental Panel on Climate Change [IPCC] 2018). Rising temperatures will affect precipitation patterns and evapotranspiration, in turn affecting all of Earth's ecosystems, including human societies. These challenges are briefly reviewed in Section 2.4. To deal with these challenges and to guarantee a more sustainable use of water resources, alternative management instruments can be used, both on the supply and demand sides. A review of these instruments is offered in Section 2.5. In this respect, an optimal water policy mix should include various instruments from both the supply and demand sides, including economic and noneconomic instruments. As water demand continues to rise, water economies are progressively reaching a maturity phase in which incremental provision costs result in an inelastic supply schedule, and the financial and environmental costs of developing new water storages have begun to exceed the economic benefits in the least productive (marginal) uses of existing supplies (Expósito and Berbel 2017; Randall 1981). At this point, the focus needs to shift from meeting demands to reallocating available resources so that the societal goals of efficiency, robustness, and resiliency are all met. From an integrated water-resource management (IWRM) perspective, this is addressed through the combination of supply-and-demand policies that ensure the best ecological status of water bodies at the least cost (United Nations [UN] 2012). Furthermore, an adequate governance framework is needed to support the implementation of the selected mix of water management instruments within a complex society with alternative (and often conflicting) interests, groups of stakeholders, and objectives. Section 2.6 aims to offers several proposals on governance issues.

2.2 WATER DEMAND: CURRENT AND FUTURE TRENDS

The water consumption worldwide has grown 800% over the last century, and for the near future, the world's population is projected to reach 9.8 billion in 2050 (UN 2018), with per capita income more than doubling by 2050. This higher income will also result in greater calorific intakes and changes in diets with an increase in the proportion of meat and dairy consumed, especially in developing countries (Alexandratos and Bruinsma 2012). The current nutrition transition in developing countries is based on the rapid growth in fat from edible plant oils and the increase in the consumption of animal products (Le Mouël et al. 2018). This combination of increased intake and dietary changes will generate an increase in demand for primary food production and, therefore, for water resources (Worldwatch Institute [WWI] 2018). Nachtergaele et al. (2011) estimate an increase in food demand (in terms of calories intake plus diet changes) in the period 2010–2050 that will require a 100% increase in crop production. The resources to cover this demand increase may come from marine, rain-fed, or irrigated agricultural areas. Nevertheless, a small increase (4%) in cultivated land is expected from 2007/2009 to 2050 (Alexandratos and Bruinsma 2012). Estimations of growth in irrigated land from 2007/2009 to 2050 is more variable and ranges from 4% (Alexandratos and Bruinsma 2012) to 11% (Nachtergaele et al. 2011). Most of the expansion of irrigated land is achieved by converting rain-fed land to irrigated land, although a minor part will take place on currently uncultivated desert land. Nevertheless, some cultivated land will be lost either as a result of deterioration or urbanization.

The share of rain-fed production in global food supply is expected to decline from 65% (2010) to 48% (2050) (Nachtergaele et al. 2011) because of the expected increase of irrigated land worldwide. This will lead to a significant increase in irrigation water withdrawals. Nachtergaele et al. (2011) estimate this increase around 11% in the period 2010–2050, and Spears (2003) considers that this could be close to 80%. The lower estimation matches the expected growth in irrigated area because intensification is not considered. Contrasting with the aforementioned increase in agricultural water withdrawal as forecast by authors specializing in the field of agricultural economics and food policy, the OECD estimates an 8% reduction in water use (2050 vs. 2010). The predictions by OECD are probably optimistic and fail to consider an increase of 11% in irrigated area for 2050 (required to satisfy increased food demand). As agricultural water withdrawal constitutes 70% of the total productive use of water resources, the precise estimate of the growth in the use of irrigation water presents a critical issue for estimation of future water demand.

Regarding the impact of climate change on water demand, it is subject to uncertainty but even if the same average precipitation is maintained, then there will be an increase in temperature, and this in turn will cause increased evapotranspiration. This effect may increase water consumption and withdrawal from agriculture over the optimistic 11% (2050 vs. 2000) predicted by FAO (Alexandratos and Bruinsma 2012), which considers that the irrigation requirement per cultivated hectare (measured in m^3/ha) remains stable. Nevertheless, we will use this "optimistic" value for our projections in the rest of this section.

In addition to agriculture, higher energy consumption, greater population, and increased urbanization will significantly increase the water demand for nonagricultural use (Expósito et al. 2019). Changes in energy consumption will be critical for water.

According to the World Energy Council (WEC 2013), energy production is expected to increase from 2010 to 2050 in the range of 27% and 61% (depending on the scenario of energy mix and prices), whereas OECD (2012) forecasts a growth around 40% (2050 vs. 2000), which implies an increase in water use for energy production of around 76%.

The growth in urbanization and per capita income will be translated into higher demand for domestic and municipal uses for horizon 2050, estimated at 183% over the 2000 levels (OECD 2012). Finally, manufacturing is expected to grow significantly because the world GDP is expected to quadruple from 2000 to 2050, and the water withdrawal for manufacturing is expected to increase 309% in this same period (OECD 2012). The combined increase of water withdrawals for all sectors is as follows: 9% for agriculture, 176% for energy, 309% for manufacture, and 83% for municipality uses; this implies a combined increase of 56% of total withdrawals (all sectors considered) in the period 2010–2050. In Table 2.1, several projections, as used by the aforementioned international institutions, have been integrated.

This global increase will exert various local impacts because more than half of the world's projected 9.8 billion people will live in water-stressed regions by 2050 (Schlosser et al. 2014). The regions that are currently under severe stress generally have a greater dependence on groundwater resources because most of them are currently overexploited (Wada and Bierkens 2014; Wada et al. 2012). Gleeson et al. (2012) estimate 1.7 billion people live in areas where groundwater resources are under threat from overexploitation.

These estimations show a pessimistic situation: already overexploited aquifers, a significant percentage of food coming from nonrenewable resources, and the fact that water withdrawal is expected to grow by around 60% by 2050. The good news is that the hydrological cycle determines that a great part of this water can be reused with reference values of return flows compared with withdrawal of around 40% for agriculture, 95% for energy (cooling), and 80% for urban and manufacturing uses (European Commission [EC] 2012). When return flows are taken into account, part of the increase in water withdrawal can be replenished with direct (wastewater reuse) or indirect water reuse (water discharge in the surface water such as river, lakes, and aquifers) that can be later reused.

TABLE 2.1
World Scenarios of Water Withdrawal by Sector

	2010 (km³/year)	2030 (km³/year)	2050 (km³/year)	Increase (%) 2050 vs. 2010
Domestic[a]	418	633	766	83
Electricity[a]	784	1167	1381	76
Agriculture[b]	2965	3097	3221	9
Manufacturing[a]	380	878	1552	309
Total	4433	5347	6920	56

Source: [a]OECD, *OECD Environmental Outlook to 2050: The consequences of Inaction*, OECD Publishing, Paris, 2012; [b]Alexandratos, N. and Bruinsma, J., *World Agriculture towards 2030/2050: The 2012 Revision*, ESA Working paper No. 12-03, FAO, Rome, 2012.

A critical question regarding water reuse is that of quality because, on a global scale, the predicted intensification of agricultural production for the growing population will drive growth in the consumption of fertilizers by around 20% on average by 2030, with India and China dominating in this respect. According to Bodirsky et al. (2014), nitrogen (N) pollution in 2050 can be expected to rise in a range of 102%–156% compared to 2010 values, meanwhile N and phosphorus (P) discharges from urban sources are estimated to multiply by a factor of 2 at global level (2050 vs. 2005) (Van Drecht et al. 2009). Furthermore, according to the UN (Ryder 2017), by 2050 1.4 billion people are projected to remain without access to basic sanitation. Even if we assume that part of these return flows can be reused, and consequently that pressure on freshwater withdrawal can be reduced, certain water users act as a critical constraint in the system. This is the case of nuclear power stations, which require a guaranteed water supply and constrain water use in the basin even when their real consumption (via evaporation) is small.

In this context of increasing water demand, the forecast increase in water withdrawals of around 60% (2050 vs. 2000) in a baseline optimistic scenario requires the implementation of innovative models of governance aimed to increase efficiency in resource use, enhance supply quality, and improve the use of an effective mix of management instruments (e.g., water pricing, market-based instruments, collective action of implied stakeholders).

2.3 SUPPLY: QUANTITY AND QUALITY CHALLENGES

According to Shiklomanov and Rodda (2004), 97.5% of the world's water is saline and only 2.5% is fresh water. Most of the world's fresh water (68.7%) is in the form of ice caps, glaciers, and permanent snow, whereas 30.1% exists as fresh groundwater. Another 0.94% is included in soil moisture, swamp water, and permafrost. This implies that only 0.26% of the total amount of fresh water on the Earth is concentrated in lakes, reservoirs, and river systems, as shown in Table 2.2.

However, the freshwater resources are unevenly distributed across the world. Renewable resources for the long term are given by the average water availability from precipitation, groundwater recharge, and surface inflows from surrounding countries. Total renewable resources in the world account for 54,453 km^3, but more than half (30,810 km^3) are located in only ten countries (Central Intelligence Agency [CIA] 2016). In 2015, 844 million people still lacked even a basic drinking water service (World Health Organization [WHO] and United Nations Childrens Fund [UNICEF] 2017). In contrast, the usage of water is higher in the United States than in any other country, with around 1582 m^3 of water withdrawals per person per year in 2013 (OECD 2016). Notwithstanding the foregoing, the scarcity or abundance of water remains a relative term. A region may have few water resources in absolute terms relative to another, but if that water is not used, there is no shortage. In contrast, there is a major scarcity if almost all available resources are used, no matter how abundant these resources may be.

Around 3800 km^3 of fresh water is withdrawn annually from the world's lakes, rivers, and aquifers. This is twice the volume extracted 50 years ago (World Commission of Dams [WCD] 2000). This, together with the intense interannual

TABLE 2.2
World's Water Resources

Water Source	Water Volume (miles³)	Water Volume (km³)	Fresh Water (%)	Total Water (%)
Oceans, seas, & bays	321,000,000	1,338,000,000	—	96.54
Ice caps, glaciers, & permanent snow	5,773,000	24,064,000	68.70	1.74
Groundwater	5,614,000	23,400,000	—	1.69
Fresh	2,526,000	10,530,000	30.10	0.76
Saline	3,088,000	12,870,000	—	0.93
Soil moisture	3,959	16,500	0.05	0.001
Ground ice & permafrost	71,970	300,000	0.86	0.022
Lakes	42,320	176,400	—	0.013
Fresh	21,830	91,000	0.26	0.007
Saline	20,490	85,400	—	0.006
Atmosphere	3,095	12,900	0.04	0.001
Swamp water	2,752	11,470	0.03	0.0008
Rivers	509	2,120	0.006	0.0002
Biological water	269	1,120	0.003	0.0001

Source: Shiklomanov, I.A. and Rodda, J.C., *World Water Resources at the Beginning of the Twenty-First Century*, Cambridge University Press, Cambridge, 2004.

and seasonal variability in many regions, makes it necessary to build dams that can ensure the availability of water for several seasons to prevent shortages. Today, 59,071 large dams (those with a height of 15 m or greater from their lowest foundation to their crest, or a dam between 5 m and 15 m impounding more than 3,000,000 m³) are registered worldwide by the International Commission on Large Dams (ICOLD; 2018) with a total storage capacity of about 16,201 km³, clearly contributing toward the efficient management of this finite resource.

However, many basins around the world have reached their maximum capacity to supply water. When a water supply falls short of its commitments to fulfill demand in terms of water quality and quantity, basins are said to be closing (Molle et al. 2010). The basin closure implies that any new demand of water must be fulfilled by decreasing the water quantity used by other users, thereby moving from supply policies to demand policies.

The risk associated to water quantity can be analyzed using an aggregate of several risk indicators, such as baseline water stress, interannual variability, seasonal variability, flood occurrence, drought severity, upstream storage, and groundwater stress (Gassert et al. 2014). Figure 2.1 shows this aggregate at basin scale and clearly presents the most threatened regions in the world, almost all of which are located within the same latitude ranges. For more information about the synthetic indicator, refer to Gassert et al. (2014).

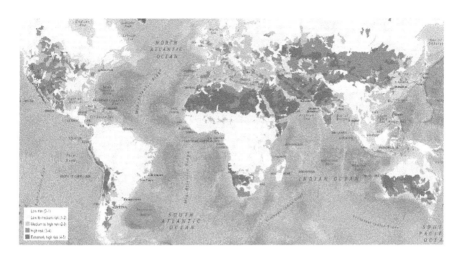

FIGURE 2.1   Water quantity risk. (From Gassert, F. et al., *Aqueduct Global Maps 2.1: Constructing Decision-Relevant Global Water Risk Indicators*, World Resources Institute, Washington, DC, 2014.)

To cope with water scarcity, new water-source technologies have emerged in many closed basins (in the sense of Molle et al. 2010) because these technologies are expensive and only appear if there is no possibility of supply expansion by increasing the number of reservoirs or extracting more groundwater. New technologies involve the replenishment of water tables, desalination of seawater, and the reuse of wastewater. Other technologies, however, are not so new or, depending on the circumstances, can they be considered as additional resources. The impacts of groundwater overexploitation are well-known and documented: they include the drying-up of qanats, springs, wetlands, and river base flows; saline intrusion along the coast and aquifers' salinization; land compaction and subsidence; dropping water tables and increased abstraction costs; as well as growing social conflicts between farmers and users (Molle et al. 2018). Groundwater is a local and highly differentiated resource, and policy instruments should be tailored to the its special hydrogeological status (Gruère and Boëdec 2019).

Managed aquifer recharge has been used for decades in many European countries, including Germany, to filtrate surface water that proceeds to the groundwater table for subsequent abstraction by wells. In Germany, managed aquifer recharge accounts for 14% of total drinking water; this percentage rises to more than 50% in Hungary and Slovakia (Sprenger et al. 2017). Although groundwater replenishment can be considered locally as a new water source, it cannot be so at basin scale unless infiltrated surface water is lost to saline sink.

A similar situation arises regarding water reuse or reclaimed water from wastewater. This source of water can be seen as a new local source from treated wastewater, but it does not constitute a new source of water at basin scale except in coastal areas, where its loss to the sea is prevented. Groundwater replenishment with treated wastewater is also common (Drewes 2009). Notwithstanding, reclaimed water can play a major role when there are no other water sources. For example, the UN Sustainable Development Goal on Water (SDG 6) specifically targets a substantial increase in recycling and safe reuse worldwide by 2030, and maximization of water reuse is a specific objective in the "Blueprint to safeguard Europe's water resources" (EC 2012). In any case, water reuse can improve the status of the environment qualitatively and can be considered a reliable water supply, totally independent from seasonal drought and weather variability (Kirhensteine et al. 2016).

Desalination of seawater can be considered as an additional source of water supply beyond the standard renewable resources stated by Shiklomanov and Rodda (2004) but is obviously relegated to coastal areas. Desalination volumes nearly doubled between 2000 and 2008 and are expected to triple by 2020. Desalinated water supply grew from 9.8 billion m^3/year in 2000 to 18.1 billion m^3/year in 2008, and it is forecast that it will reach 54 billion m^3/year in 2020 (Sommariva 2017). Several economies in the Arabic Gulf rely on desalination to produce 90% or more of their drinking water, and the overall capacity installed in this region amounts to about 40% of the world's desalinated water capacity. Around 60% of desalination capacity treats seawater; the remainder treats brackish and less saline feedwater (Sommariva 2017).

Thermal desalination (distillation) has been progressively substituted by more widely used and less energy-dependent semipermeable membrane processes (mainly reverse osmosis, but also electrodialysis and membrane distillation). Much progress has been made in desalination technology, although its cost remains too high for its general use in agriculture, and it is therefore mainly used by households and industry. Its high cost is because of the investment and the amount of energy required by membrane processes, despite costing less than thermal desalination. In consequence, the cost of desalination, depending on site- and country-specific factor ranges between $1 and $4 per cubic meter (Buros 2000; Sommariva 2017). Certain promising technologies include its combination with renewable energy sources (Mathioulakis et al. 2007), forward osmosis (Nicoll 2013), and other alternatives briefly summarized in Miller (2003), which include solar stills, freeze desalination, or other crystallization processes, membrane distillation, and wave-driven processes.

Water quality constitutes another major issue because of its impact on the environment and human welfare. Water quality must be understood as the ecological status of water bodies as a whole. For example, the most widespread pressures on ecological status in the European Union (EU; 19 member states) are those of hydromorphological nature, which affect about 40% of the water bodies (EC 2012). Chemical quality and quantity are also related; chemical quality indicators always refer to a volume of water, and hence changes in water quantity result in changes in water chemical quality. In general, countries or regions with abundant water are more concerned with qualitative aspects, whereas countries or regions with relative water scarcity are more concerned with the amount available. The latter, however, are also becoming increasingly concerned with the quality aspect and the former are becoming increasingly concerned with the quantity aspect because of the intensification in water use and to climate change. Environmentally concerned governments have enacted legislation to protect their bodies of water. Good examples are provided by the US Clean Water Act and Safe Drinking Water Act to maintain the chemical, biological, and physical integrity of water bodies and to protect public health through regulating the nation's public drinking water supply and by the European Water Framework Directive enacted for the protection of inland surface waters, transitional waters, coastal waters, and groundwater.

Water pollution is classified as "point source pollution" when it refers to wastewater directly discharged into stream flows (typically from households or industry), and "nonpoint source pollution" refers to pollution from diffuse sources, such as agricultural runoff into rivers or percolation into aquifers. Wastewater treatment is the common way to cope with point source pollution and is regulated by law in most countries. The main contaminants from point and nonpoint source pollution are nitrogen and phosphorus, which lead to the eutrophication of water bodies, resulting in hypoxia with detrimental consequences for fish and aquatic invertebrates (Vollenweider 1970). Integrated pest management can also reduce the pollution from pesticides. Runoff with sediment and pollutants from agricultural areas can be reduced by applying a series of common techniques including contour ploughing, crop mulching, crop rotation, planting perennial crops, installing riparian buffers, and using conservation tillage. Additionally, a better management of commercial fertilizers and extensive livestock farming can diminish nutrient concentration in the runoff. Better water-use efficiency (drip sprinklers) can also significantly reduce the return flows from agriculture, thereby diminishing the pollutants in water bodies (Lecina et al. 2010).

2.4   CLIMATE CHANGE, FLOODS, AND DROUGHTS

Among the many global challenges affecting water resources from both supply and demand sides, climate change occupies a principal position. It is estimated that human activities have already caused a 1.0°C increase in global temperature levels as compared to preindustrial levels, with a confidence interval between 0.8°C and 1.2°C. After 2030 and before 2052, global warming is likely to reach a 1.5°C increase unless decisive action is taken by governments worldwide (IPCC 2018). Rising temperatures will affect precipitation patterns, although there is significant uncertainty involved (IPCC 2014): with high confidence, the mean precipitation over mid-latitudinal land areas in the Northern Hemisphere has increased since 1951; elsewhere, long-term increases and decreases mean precipitation trends are of low confidence. For heavy precipitation events, IPCC (2014) found a larger number of areas with more increases than decreases in the frequency and intensity of the events, which suggest a "global-scale intensification of heavy precipitation" (medium confidence) (IPCC 2018). Regarding droughts, there is low confidence regarding global-scale trends, though robust regional trends are observed in the Mediterranean and West Africa (drought increases) and central North America and northwest Australia (drought decreases) (IPCC 2014). In the Mediterranean area, recent literature suggests that drought likelihood has substantially increased and that the (already growing) drought trends available are worsening and should be revised (medium confidence).

At 1.5°C and 2°C increase in global warming, projected changes in regional precipitation show "robust differences in mean precipitation compared to the pre-industrial period," with reductions expected in the Mediterranean area, the Arabian,

Peninsula and Egypt and increasing mean precipitation trends in high latitudes (IPCC 2018). Regarding heavy precipitation, experts also observe robust changes at both 1.5°C and 2°C as compared to preindustrial levels. In Europe for instance, there is agreement in the existence of a positive change in heavy precipitation at 1.5°C. Projections of changes in drought and dryness are uncertain in many regions, although the Mediterranean, US Southwest, and southern African regions display consistent dryness and growing drought trends in most assessments.

Climate change can exacerbate or offset water availability and water security problems worldwide, which appear, nonetheless, to be dominated by population and socioeconomic trends. Economic losses from flooding since 1951 have increased mainly because of greater exposure and vulnerability (high confidence), and impacts of drought have mostly increased because of growing demand (IPCC 2014). At 1.5°C and 2°C increase in global warming, risks associated with runoff (e.g., landslides) will increase globally, and flood hazards will also increase in certain regions, although socioeconomic conditions will become a more relevant factor in explaining socioeconomic losses (Alfieri et al. 2017). However, assuming constant populations, countries representing 73% of the world's population would still face an increasing flood risk of between 100% (1.5°C) and 170% (2°C). Under "constant socio-economic conditions," the population exposed to extreme drought at 1.5°C in 2021–2040 is expected to be 114.3 million, and 190.4 million at 2°C in 2041–2060.

Response to droughts used to be "reactive," managing a water crisis by measures after a drought event takes place. During the last two decades, there has been an increasing shift toward a "proactive" approach. The proactive or preventive approach that can be considered an approach to "risk management" consists of measures to be prepared in a planned manner in case of drought. The first drought management plans were promoted in the Mediterranean region, where we can find several guides for the preparation of such plans (Iglesias et al. 2007; Spanish Ministry of Environment 2005). An example of a result of these initiatives, the drought management plans of Spain arose in 2007 as part of the River Basin Management Plans (Estrela and Vargas 2012), which served as an example for other plans, such as those of Turkey or the United States.

Finally, regarding flood risk management, there is growing literature on this issue and several models for predicting floods have been developed in recent years, as well as methods to evaluate different technical and nontechnical measures to manage flood risks. These methods include planting vegetation to retain extra water, terracing hillsides to arrest downhill flow, and the construction of floodways (man-made channels to divert floodwater). Other techniques include the construction of levees, dikes, dams, reservoirs, or retention ponds to hold extra water during times of flooding (International Water Association [IWA] 2016).

2.5   WATER MANAGEMENT INSTRUMENTS FOR THE FUTURE

Demand pressures, supply constraints, and consequences of climate change all act to increase competition for scarce water resources and make the (re)allocation of water resources a rising issue in the policy agenda (Damania et al. 2017). Water-resource (re)allocation refers to the distribution of the resource across space and time and among users (OECD 2015a). Policy makers have several instruments at their disposal to encourage an efficient allocation system that represents society's preferences and to ensure this system is robust and resilient and thus capable of tolerating and recovering from perturbations. Traditionally, such instruments have been based on engineering, that is, focused on the construction and exploitation of waterworks to meet water demand and harness the potential of water for economic growth. From aqueducts, reservoirs, and traditional irrigation systems, waterworks have escalated to interbasin water transfers, major dams, modern irrigation devices, wastewater treatment plants, and desalination plants, among others (Hassan 2010). Yet, as water demand continues to rise, water economies are progressively reaching a maturity phase in which incremental provisioning costs result in an inelastic supply schedule, and the financial and environmental costs of developing new waterworks have begun to exceed the economic benefits in the least productive (marginal) uses of existing supplies (Randall 1981). At this point, the focus needs to shift from meeting demand needs to reallocate available resources so that the societal goals of efficiency, robustness, and resiliency are met. From an IRWM perspective, this is addressed through the combination of supply and demand policies that "maximize the resultant economic and social welfare in an equitable manner without compromising the sustainability of vital ecosystems" (UN 2012). Demand-side policies typically involve regulatory instruments (command and control) that specify a particular type of behavior agents have to comply with but also include economic instruments that replace the traditional notions of control and government-led planning by those of incentives, motivation, and multilevel governance. This section surveys supply- and demand-side instruments for water resources management and explores their pros and cons, including issues of socioeconomic context, policy mix, and sequencing, so to assess their contribution to sustainable, robust, and resilient economic growth.

2.5.1 WATERWORKS

John Locke and Francis Bacon famously stated that "nature is only subdued by submission." Throughout human history, water policies worldwide have reproduced this view. Following a nomadic period where hunters and gatherers depended on the wild plants and animals sustained by rainfall (which varied significantly from one place to another but, on the whole, was insufficient to provide food for large, dense, settled populations), families began settling near springs, lakes, and rivers to supply livestock and crops with water, gradually developing technologies to divert water for irrigation and domestic purposes. Many civilizations, from Babylonian to Chinese, Mayan, or Roman, constructed water delivery systems such as aqueducts to carry water to cities (Hassan 2010; Yevjevich 1992). These and other settled societies thereafter have addressed water quality and quantity issues through capital-intensive waterworks that made increasing amounts of water available to users. These include reservoirs, canals, wells, water transfers, irrigation schemes and, more recently, desalination and wastewater treatment plants. Recent paradigmatic examples of this trend can be found in the arid and semi-arid areas of the Mediterranean basin, southern Australia, Chile, and the western states of the United States, where economic growth has been closely linked to the capacity of public institutions to make increasing amounts of water available to users. In this context, the main objective of water policy consisted of finding inexpensive and reliable means to meet water demand through a coordinated public effort to supply the water services demanded as a result of advances in the many areas of the economy, including population growth, urban sprawl, expanding manufacturing activities, and irrigation development.

However, this technical success in harnessing the potential of water for economic growth has come with a number of problems. Failure to acknowledge nature's limits and the finiteness of water resources has led to unrealistic expectations of the ability of water bodies to meet growing needs from human systems, which in turn have resulted in increasing demand and water-resource overallocation. Slowly but steadily the law of diminishing returns, which states that increasing the amount of a single factor of production (in this case, water) leads to a decrease in the marginal incremental output of the production process, has eroded marginal returns of water projects. A good example to illustrate this trend can be found in hydropower in Spain, where the installed capacity has increased by 145% from 1928 to 2012, but the average hydropower production (in GWh) throughout the period remained at similar levels to those of 1928. Although admittedly this policy increased the robustness and resilience of the system, its financial costs are considerably larger than these benefits, and the cost-to-benefit ratio worsens when we include environmental costs (Expósito 2018; Gómez 2009). Supply-side policies represent a cornerstone in the development of human societies. However, the water crisis we are facing cannot be tackled solely with waterworks; demand instruments are necessary to define priorities and reallocate *available* resources so that societal goals are met (Quiggin 2001).

2.5.2 REGULATORY INSTRUMENTS

In the last few decades, water-resource management has become a problem that can no longer be treated exclusively from the technical standpoint (i.e., through water-supply policies). The unparalleled increase in water withdrawals has forced policy makers to implement several measures also concerning the demand for water. This has been carried out, to a large extent, through a more intensive use of command-and-control policies. Command-and-control policies are not new in water policy, and they have been used as a complement to supply policies for centuries. However, more recently they have evolved from simple rules, which were only casually enforced by law, to highly complex and sophisticated water management plans, that are fully supported by local, regional, national, or supranational (formal and informal) institutions (Pahl-Wostl et al. 2010).

Command-and-control refers to a set of regulatory instruments that specify a particular type of behavior with which agents have to comply. Traditionally, the intended behavior was decided unilaterally and enforced by legal disciplining using the power of the state. More recently, many institutions worldwide have opted for a more integrative approach in which the objectives of the command-and-control policies are designed through a social agreement including all the agents affected, where decisions are ultimately enforced by formal (e.g., the public sector) but also informal institutions (Santato et al. 2016).

From the perspective of an omniscient state with the capacity to enforce water restrictions, command-and-control policies are sufficient to achieve the objectives of water policy. However, the institutional design and implementation of these regulations present nontrivial challenges in real life. The cost of enforcing command-and-control policies depends on the economic incentives toward compliance (or penalties for non-compliance) and on the complexity of the command-and-control policy itself. The latter is of special relevance in the case of water because the status of water resources is the result of the action of multiple agents, which makes it complex to command and control all the factors involved to achieve the desired societal objectives. For example, the application of law-enforcement mechanisms requires structures to inspect and apply fines and penalties, with increasing difficulties because of the magnitude of the problem. This is particularly true when the system is operating under critical situations of water stress (e.g., multiple sources of pollution, overexploitation, etc.) that increase the marginal value of water and strengthen the incentives toward non-compliance. The challenges associated with the implementation of conventional command-and-control policies have made researchers argue in favor of a complementary incentive-based water management approach (Gómez et al. 2017). Note that both supply policies and economic incentives are inherently linked to command-and-control policies, which are a prerequisite both to define a common legal framework in which agents interact and to set general guidelines in water policy for the achievement of collectively agreed societal goals.

2.5.3 ECONOMIC INSTRUMENTS

The current water crisis is now recognized as being largely a crisis of governance and not of resources or technological constraints (Bucknall 2006; Hanemann 2006). Although the technical capacity of society to put additional amounts of water at the service of growing water demand has increased exponentially, considerations of a social, political, institutional, and financial order are still treated in an incipient form with major problems persisting (Pahl-Wostl et al. 2013). This failure is exemplified by the inability of the current policy mix to match the decisions made by water users in the economy with the ability of the existing water resources to satisfy these uses in a sustainable manner.

Economic instruments can contribute toward addressing this problem by substituting the traditional notions of control and government-led planning with those of incentives, motivation, and multilevel governance. However, evidence demonstrates that there is still a major gap between the political rhetoric and the operational level. With a few exceptions (see, e.g., the case of the German sewage water levy) (Gawel and Faelsch 2012), economic instruments for water management are seldom found outside academic papers, and in many cases, they consist of financial tools disguised as economic instruments to make the ultimate goal of raising revenues more acceptable (Gómez et al. 2017).

As a result of the little evidence available and of the misuse of economic instruments, there is significant confusion regarding what an economic instrument for water management actually is. Although it is generally accepted that taxes, fees, subsidies, and markets can all be considered economic instruments, there is still disagreement regarding the inclusion of nonmarket mechanisms in this group and the purpose and design of economic instruments. This confusion is perceivable even in the academia. Various definitions have been made available, and in certain aspects, they conflict with each other. For example, the National Center on Education and the Economy (NCEE 2001) considers that economic instruments are financial tools (i.e., market based) that "provide monetary and near-monetary rewards" for the accomplishment of environmental goals. In the same respect, Stavins (2003) puts economic instruments at the same level as market-based instruments and labels them as "harnessing market forces." Kraemer et al. (2003) open up the definition to nonmarket economic instruments, but they fail to distinguish between economic instruments for water management and those for raising revenue: "When the primary aim of an environmental charge or tax is not to create incentives but to raise revenue, the relevant distinction lies in whether the revenue is earmarked or simply added to the general government budget." Although financial instruments are of paramount importance for the accomplishment of water policy and societal goals, they should not be regarded as economic instruments for water management where they do not target effectively water policy objectives. The economic rationale for this is provided by the Tinbergen and Assignment Principles. The Tinbergen Principle, developed by the first Nobel Prize in Economics, Jan Tinbergen, states that the attainment of a number of targets necessitates an equal number of instruments (Tinbergen 1952); an essential complement to Tinbergen's research comes from another Nobel Prize, Robert Mundell, who in his Assignment Principle states that each instrument should pursue a single target and never be used to pursue a second target (Mundell 1962).

All the definitions highlight that the key features of economic instruments are those of incentives, motivation, and voluntary choice. They also stress that at least one of their objectives should be that of adapting individual decisions to collectively agreed environmental goals. Strosser et al. (2013) gather up all these contributions and define economic instruments as "those incentives designed and implemented with the purpose of adapting individual decisions to collectively agreed goals." This implies that financial instruments aimed primarily at raising revenue are not economic instruments for water management (although cost recovery can be a secondary objective of these); on the other hand, cooperative agreements and other nonmarket instruments that lead to behavioral changes may be economic instruments for water management even in the absence of pecuniary transactions. An illustrative example could be that of urban water levies because urban water demand is often inelastic, a tariff is unlikely to reduce water use significantly but can become a relevant source of tax revenue. If the policy goal is to restore the balance in an overallocated basin, this instrument should not be regarded as an economic instrument for water management; it is a revenue-raising tool. Recent reviews on the characteristics, advantages, and disadvantages of economic instruments are available, for example, in Gómez et al. (2017) and Rey et al. (2018). A summary is provided in Table 2.3.

It should be noted that economic instruments for water management are not a substitute for conventional command-and-control and supply policies; rather, they should be designed to complement them. This presents the challenge of singling out the actual contribution of an economic instrument toward the goals of the water

TABLE 2.3
Economic Instruments for Water Resources Management

Economic Instrument	Application to Water Resources Management	Pros	Cons
Subsidies	Subsidies for water resources management are typically directed to encourage the provision of positive externalities through, for example, adoption of desirable technologies toward flood prevention (e.g., levees) and water conservation (albeit with controversial outcomes, see, e.g., Perry and Steduto (2017). Subsidies can be explicit (including price support, subsidized loans, and direct payments) or implicit (including tax or charges relief and reduced regulation).	Acceptability, limited transaction costs; decoupled subsidies do not constitute an economic instrument as per the definition but can complement other economic instruments (e.g., equity issues).	Budgetary pressures may lead to economic inefficiencies (e.g., rebound effect in irrigation modernization), may encourage rent-seeking behavior.
Water charges	From a conventional economic standpoint, the optimal water management instrument is charging the resource at the margin in accordance with the social cost of water (i.e., private cost plus net externalities). Such mechanisms would make any other economic instrument unnecessary and distortive. However, the political economy of water resources means that socially efficient charging can be "impractical on technical and political grounds" (Berbel et al. 2019; Fishman et al. 2015).	Effectiveness; adequate price signals do reflect actual resource costs and avoid over- or underpricing; encourages efficient water (re)allocation; encourages adoption of new, potentially more efficient technologies; flexibility (e.g., incremental charging); tax revenue.	Requires knowledge of price elasticity to determine the adequate charge and avoid over- or underpricing; require a census of water users and rights and water abstraction metering (which may be unfeasible, e.g., in rice systems water flows from field to field and through the saturated zone) or an alternative technique to determine or limit water use (e.g., indirect groundwater charging via energy charging); asymmetric impacts and inequality may demand complementary compensatory measures (e.g., decoupled subsidies).

(Continued)

TABLE 2.3 (*Continued*)
Economic Instruments for Water Resources Management

Economic Instrument	Application to Water Resources Management	Pros	Cons
Payment for ecosystem services (PES)	PES are a category of economic incentive approaches where the user or beneficiary of watershed services exchanges financial value for the provision of protection, rehabilitation, or enhancement of watershed services or land uses from sellers or providers (Asbjornsen et al. 2015).	Flexibility and acceptability; increase the revenue of land owners by securing or increasing the production of environmental services; contribute to reinforce the political voice and legitimacy of stakeholders during negotiation process; increase the provision of target environment services and complementary ones.	Budgetary pressures; inefficiency due to lack of adequate performance monitoring; free riding associated to the nature and functioning of ecosystem services; infringement of the polluter-pays principle (e.g., European Union).
Water markets	A market for water "permits the temporary, long-term, or permanent transfer of water from the existing rights-holders to other water users in exchange for payment" (Hanak 2003, p. 2). Water trading will only occur if there is a difference, after transaction, transport, and risk costs, between buyer's willingness to pay and a seller's willingness to accept payment for not having that water available (Calatrava and Garrido 2005).	Creates market incentives toward allocative improvements; can deal with externalities through public sector engagement (e.g., buyback); reveals water users the opportunity costs of the resource.	Acceptability; transaction costs; institutional complexity; social disparities and reducing spatial cohesion as water is (re)allocated to more valuable uses; speculation with water rights that are accumulated and not used; worsening overexploitation and scarcity trends if water use rights do not match available water resources; incentives toward higher consumption-to-withdrawal ratios and depletion as water is transferred to more efficient users.

(Continued)

TABLE 2.3 (*Continued*)
Economic Instruments for Water Resources Management

Economic Instrument	Application to Water Resources Management	Pros	Cons
Insurance	Through insurance schemes, the policyholder "transfers the cost of potential loss to the insurer in exchange for monetary compensation known as a premium" (Rey et al. 2018).	Alternative way to stabilize farmers' income during dry periods to (e.g., unsustainable aquifer overdraft); creates conditions for collective control of aquifers; privately funded, albeit public subsidization is common, particularly in agricultural insurance (i.e., limited budgetary pressures).	Willingness and ability to pay; institutional and legal complexity; public subsidization can be costly and lead to fiscal imbalances; over- and undersubsidization can lead to equity issues.
Voluntary agreements	Nonpecuniary and voluntary agreements among the government, businesses, individuals, or nonprofit organizations toward improving water resources management based on opportunities for profit or loss mitigation.	Flexible; effective; acceptable; inexpensive.	Difficulty to find win-win opportunities for voluntary agreements often result in lack of consensus; free riding and problems with implementation and enforcement.

policy: economic instruments are never implemented in isolation from other supply and command-and-control policies (nor should they be); and many macroeconomic and sectoral changes that influence behavior and water use also take place at the same time. Beyond the selected economic instrument itself, focus should be placed on issues regarding its sequencing, the adequate mix and number of instruments in relation to the objectives, and the socioeconomic and environmental contexts. For example, in Australian water markets, one can easily differentiate between the disappointing environmental performance of private markets and the promising outcomes achieved through public buyback programs. Water markets and other economic instruments are not a *panacea* for water management problems; instead, they are creatures of design whose performance depends on critical design features, such as the context (i.e., hydrologic, institutional, socioeconomic, and policy), policy mix, and sequencing.

2.6 CHALLENGES IN WATER GOVERNANCE AND MANAGEMENT

Water management instruments constitute the technical (or practical) side of a more general policy framework for water governance. The concept of governance is normally associated to the human component of water management (Cosgrove and Loucks 2015) in contrast to the technical component based on either the development of management instruments to obtain certain results (Linton and Brooks 2011) or, similarly, on specific management actions to achieve certain objectives (Pahl-Wostl 2007). Water governance could be thus understood as the way human decisions are made to facilitate water management aimed at achieving certain goals. The multiple uses of water resources (e.g., agricultural, industrial, and urban uses) and the competition of provision consumptive users with environmental services and their intrinsic characteristics (e.g., common-pooled resource, public goods) require the implementation of adequate governance frameworks to guarantee a sustainable and efficient management of this resource. Additionally, the water sector is characterized by the existence of market failures, such as the high-capital intensity and monopolistic nature of surface-water management and utilities. This requires the development of adequate institutional and legal frameworks to guarantee an efficient, effective, and socially equitable management of water resources. These special characteristics make water governance extraordinarily challenging because it should involve the management of all water resources (i.e., surface water and groundwater, including coastal waters) in a multilevel framework (e.g., national, regional, local) with the participation of private and public stakeholders throughout the whole process: decision making, management and policy design, and implementation and evaluation processes. Additionally, ecological aspects need to be covered because water bodies constitute a primary source of ecosystem services. Furthermore, water governance is inherently complex because of its direct connections to the development sphere of countries and regions, such as agriculture, environment, energy, health, poverty, etc. This complexity normally requires the distribution of policy and management responsibilities among different spatial levels (e.g., national, regional, local, and watershed), although this may generate policy fragmentation (OECD 2011, 2015b).

Although there is no one-size-fits-all solution for water governance that is capable of facing future challenges worldwide because governance frameworks are highly context-dependent and decisions should be adapted to spatial specificities, water governance should be understood as "a means to an end" rather than "an end in itself." The OECD has provided evidence on the main governance gaps that hinder the design and implementation of water policies in the last two decades and propose strategies for water policy reform. In doing so, it suggests an analytical framework to identify and bridge governance challenges that affect, to a greater or lesser extent, all countries, regardless of their institutional setting, water availability, and degree of decentralization (OECD 2011). This framework is summarized in Figure 2.2 and is based on three mutually reinforcing and complementary dimensions of water governance: effectiveness, efficiency, and trust and engagement. Effectiveness relates to the contribution of governance toward defining clear sustainable water policy goals and is aimed at all levels of government for the implementation of specific policy goals and attainment of expected targets. In this governance dimension, it is crucial to define clear roles and responsibilities of all the decision makers involved in water governance (public and private) on all decision-making scales (i.e., national, regional, and local).

Governance must be defined at proper scale, and according to many institutions such as the EU, the most relevant administrative boundary is the basin or aquifer (as specified by the EU Water Framework Directive), although certain decisions and management practices are carried out at national, regional, and local levels (Berbel and Expósito 2018). Independent of the established management scales, people

FIGURE 2.2 Water Governance Principles. (From Organisation for Economic Co-operation and Development (OECD), *Principles on Water Governance*, Adopted by the OECD Regional Development Policy Committee on May 11, 2015. Available at http://www.oecd.org/cfe/regional-policy/OECD-Principles-on-Water-Governance.pdf, 2015b.)

involved in decision making (e.g., policy makers and technical and administrative staff) should be armed with the knowledge and capacity to make coherent policy decisions in line with those other decisions taken on other management scales. Efficiency relates to the contribution of governance toward maximizing the benefits of sustainable water management and welfare at the least cost to society. This principle requires the development of innovative governance initiatives based on the use of available data and information in the water sector, as well as on the development of appropriate regulatory frameworks capable of adapting to continuously changing conditions in the water sector. To acquire the information necessary for the design of efficient (and effective) water policies, the implementation of adequate and sufficient monitoring measures is crucial in both water quantity and quality dimensions. Furthermore, the water sector heavily relies on financial resources to finance supply policies when needed, such as the construction of water infrastructures and also for the development of demand-management instruments, such as the implementation of ecosystem services and market-based solutions. Finally, the trust and engagement dimension relates to the contribution of governance toward building public confidence and ensuring the inclusion of all stakeholders involved in the water sector (including the general public). With this aim, effective stakeholder engagement requires integrity and transparency in policy making to build mutual trust and confidence. Furthermore, the recognition of trade-offs across stakeholders, both in current and future scenarios, remain necessary for social agreements to be reached. In this respect, the recognition of environmental, cultural and, in general, nonuse economic values of water resources (such as existence, altruistic, and legacy values) play a significant role. At the same time, these social agreements (based on trust and engagement) must be monitored and evaluated in a continuous process because conditions may change rapidly as a result of any of a multiple of factors (e.g., climate change, stakeholders' generation renewals, new legislation frameworks, etc.).

In line with the OECD's water governance principles, the approach of IWRM has been accepted internationally as a suitable tool for the efficient, equitable, and sustainable development and management of the world's limited water resources and for coping with conflicting demands, thus going beyond simple resource management to guide policy and decision making (governance dimension). The Global Water Partnership defines IWRM as a "process which promotes the coordinated development and management of water, land and related resources, in order to maximize the resultant economic and social welfare in an equitable manner without compromising the sustainability of vital ecosystems" (GWP 2000, p. 22). Thus, IWRM is a comprehensive, multidisciplinary, and cross-sectoral planning and implementation tool for managing and developing water resources (UNEP 2009). Following Aldaya and Ramón-Llamas (2017), IWRM involves the collection and management of information on natural resources, the understanding of the interactions of its use, and the implementation of policies, practices, and administration structures to enable these resources to be used.

The complexity of water governance increases in the case of transboundary water bodies because its management usually requires the development of transboundary institutions and international cooperation frameworks for decision making. To date, 263 transboundary lake and river basins and 592 transboundary aquifers have been identified (UN Environment Programme [UNEP] 2016). Managing transboundary water bodies requires building trust and effective cooperation between countries, and despite progress having been reported in certain regions (e.g., Europe and North America), transboundary governance efforts and management initiatives appear to remain insufficient. In this respect, the Transboundary Water Assessment Programme (TWAP) implemented by UNEP (2016) started in 2013 with two major objectives: (i) to carry out the first global-scale assessment of transboundary water systems that will assist international organizations to improve transboundary governance and minimize conflicts and (ii) to ensure that transboundary considerations are incorporated into the international political agenda.

In the global context of increasing water demand pressures and rising resource scarcity, international conflicts are more likely to occur when the presence and capacity of the institutional mechanism (e.g., treaty, institution) are insufficient to address changes in the social, economic, or environmental conditions of the water body (Yoffe et al. 2003). In this regard, the development of effective transboundary governance measures is urgent because of the growing water challenges that humanity will have to address in the future. Global-water governance needs the involvement of not only governments but also nonstate stakeholders (e.g., companies, civil society) to prevent water crises rather than adapting in response to them (Cosgrove and Loucks 2015). Additionally, instruments of international water laws must be developed urgently with the aim of introducing the necessary adaptability of international cooperation in water resources worldwide in the context of climate change and increasing occurrence of extreme events (e.g., droughts and floods) (IWA 2016).

In summary, the design of an adequate water governance framework (including the needed institutional and legal reforms at regional, national, and international levels) together with an optimal mix of water policy and management instruments (as discussed in Section 2.5) emerges as the only possible solution to guarantee a sustainable management of water resources. Furthermore, this solution should be tailored to specific countries and regions (e.g., river basins), whereby their specific socioeconomic, agronomic, ecosystem, and climatic conditions are considered.

2.7 CONCLUSION

The set of the Millennium Development Goals (MDGs), which expired at the end of 2015, constituted the first attempt to set targets for development goals on a global scale. Nevertheless, not all global water challenges were addressed (e.g., climate change and water-scarcity challenges). Specifically, the target to halve the proportion of the population without sustainable access to safe drinking water and basic sanitation by 2015 has been achieved (2.6 billion people with improved access to water services). The SDGs adopted by the UN General Assembly in 2015 follow up on the MDGs with the horizon set at 2030. The seventeen goals and more than 160 targets encompass a wide variety of relevant issues for the planet related to the use of water resources and environments, including the provision of clean water and sanitation (goal 6), actions to combat climate change (goal 13), protection of the oceans (goal 14), and the sustainable use of natural resources (goal 15). For the first time in human history, these goals take into account the need to preserve water sources, both in terms of quantity and quality, through the promotion of a sustainable use of these resources to ensure adequate levels of well-being for current and future generations.

The challenge of making population growth compatible with increased living standards and per capita income, and with decreased malnutrition (as included in almost all SDGs) is affected by water scarcity and poor water quality. This chapter has attempted to integrate the forecast of demand and supply, which will define a likely scenario for 2050, and has found that the need for an approximately 60% increase in water withdrawal would increase pressure on the environment and would need to be mitigated with the support of a set of water-related technical innovations (e.g., desalination, enhanced water-use efficiency) and innovations from other sectors (e.g., energy efficiency, increased crop productivity). However, increased supply and efficiency will be insufficient to match supply and demand; hence, economic instruments (e.g., water markets, water pricing) and governance institutions will be required to guarantee that economic, social, and environmental goals are achieved.

Because water demand is a mainly a derived demand (i.e., irrigation water is an input of food production, and cooling water is an input for energy production), the increased scarcity of water and its increased cost will induce changes in the demand for final goods. The fact that agriculture will account for 47% of water withdrawal, according to the projections in Table 2.1, and that the agricultural share of water consumption may stand at around 83% of global-water consumption (net of return flows) indicates that an increased cost for water may lead to changes in agricultural prices and thus, in the final demand for food. This will also be a compensating force toward facilitating the adaptation to the growing global water scarcity, as this chapter reveals.

This chapter has aimed to offer an overview on where we stand regarding the use of water resources on a global scale and the challenges to be faced in the near future. Growing demand pressures, increasing water scarcity in terms of quantity and quality, the more evident effects of climate change on water resources, and the challenges at achieving an effective and sustainable water management and governance at local, national, and international levels constitute the significant factors to be taken into account to guarantee the sustainable use of water resources for the present and future generations.

A suitable water governance framework with the use of an optimal mix of water policy and management instruments, together with feasible institutional reforms and the necessary investment, emerges as the only possible solution. This solution should be tailored to specific countries and regions (e.g., river basins), whereby their specific socioeconomic, agronomic, and climatic conditions are considered. Global challenges related to the long-term sustainable use of water resources are extremely difficult to address because they require time, money, and political commitment— elements that are seldom applicable in the long term. In this context, the participation of all involved stakeholders becomes a crucial factor for consideration in the development of a sustainable governance framework of water resources. We believe that decision makers will act with due diligence because time is running out, water resources are limited, and challenges thereof can only increase.

REFERENCES

Aldaya, M., and M. R. Llamas. 2017. Towards an Integrated Water Resource Management (IWRM). In *Water, agriculture and the environment in Spain: Can we square the cycle?*, ed. L. De Stefano and M. R. Llamas, 67–73. Boca Raton, FL: CRC Press/Balkema.

Alexandratos, N., and J. Bruinsma. 2012. *World agriculture towards 2030/2050: The 2012 revision*, ESA Working paper No. 12-03, Rome: FAO.

Alfieri, L., Bisselink, B., Dottori, F., Naumann, G., Roo, A. de, Salamon, P., Wyser, K., and L. Feyen. 2017. Global projections of river flood risk in a warmer world. *Earths Future* 5: 171–182.

Asbjornsen, H., Mayer, A. S., Jones, K. W., Selfa, T., Saenz, L., Kolka, R. K., and K. E. Halvorsen. 2015. Assessing impacts of payments for watershed services on sustainability in coupled human and natural systems. *Bio Science* 65: 579–591.

Berbel, J., and Expósito, A. 2018. Economic challenges in the EU Water Directive Framework reform and implementation. *Eur Plan Stud* 16(1): 20–34.

Berbel, J., Borrego-Marín, M., Expósito, A., Giannoccaro, G., Montilla-López, N., and C. Roseta-Palma. 2019. Analysis of irrigation water tariffs and taxes in Europe. *Water Policy* 21: 806–825.

Bodirsky, B. L., Popp, A., Lotze-Campen, H., Dietrich, J. P., Rolinski, S., Weindl, I., Schmitz, C. et al. 2014. Reactive nitrogen requirements to feed the world in 2050 and potential to mitigate nitrogen pollution. *Nat Commun* 5: 3858.

Bucknall, J. 2006. *Good governance for good water management*. Environment Matters. Annual Review. Washington, DC: World Bank Publications.

Buros, O. 2000. *The ABCs of desalting*. Topsfield, MA: International Desalination Association.

Calatrava, J., and A. Garrido. 2005. Spot water markets and risk in water supply. *Agric Econ* 33(2): 131–143.

Central Intelligence Agency (CIA). 2016. *The World Factbook 2016–17*. Washington, DC: CIA.

Cosgrove, W. J., and D. P. Loucks. 2015. Water management: Current and future challenges and research directions. *Water Resour Res* 51: 4823–4839.

Damania, R., Desbureaux, S., Hyland, M., Islam, A., Moore, S., Rodella, A.-S., Russ, J., and E. Zaveri. 2017. *Uncharted waters: The new economics of water scarcity and variability*. Washington, DC: World Bank Publications.

Drewes, J. E. 2009. Ground water replenishment with recycled water—Water quality improvements during managed aquifer recharge. *Groundwater* 47: 502–505.

Estrela, T., and E. Vargas. 2012. Drought management plans in the European Union: The case of Spain. *Water Resour Manag* 26: 1537–1553.

European Commission (EC). 2012. *A blueprint to safeguard Europe's water resources*. Brussels: EC.

Expósito, A. 2018. Irrigated agriculture and the cost recovery principle of water services: Assessment and discussion of the case of the Guadalquivir River Basin (Spain). *Water* 10(10): 1338.

Expósito, A., and J. Berbel. 2017. Agricultural irrigation water use in a closed basin and the impacts on water productivity: The case of the Guadalquivir river basin. *Water* 9(2): 136.

Expósito, A., Pablo-Romero, M., and A. Sánchez-Braza. 2019. Testing EKC for urban water use: Empirical evidence at River Basin Scale from the Guadalquivir River, Spain. *J Water Res Plan Manag* 145(4): 04019005.

Fishman, R., Devineni, N., and S. Raman. 2015. Can improved agricultural water use efficiency save India's groundwater? *Environ Res Lett* 10: 084022.

Gassert, F., Luck, M., Landis, M., Reig, P., and T. Shiao. 2014. *Aqueduct Global Maps 2.1: Constructing decision-relevant global water risk indicators*. Washington, DC: World Resources Institute.

Gawel, E., and Faelsch, M. 2012. On the steering effect of Wastewater Levies Part 1: Steering functions and substitution effects. *KA- Korrespondenz Abwasser*, Abfall 59(11): 1060–1065.

Gleeson, T., Wada, Y., Bierkens, M. F. P., and L. P. H. van Beek. 2012. Water balance of global aquifers revealed by groundwater footprint. *Nature* 488: 197–200.

Global Water Partnership (GWP). 2000. Towards water security: A framework for action. Global Water Partnership, Stockholm, Sweden.

Gómez, C. M. 2009. La eficiencia en la asignación del agua: Principios básicos y hechos estilizados en España. *Inf Comer Esp Rev Econ* 847: 23–39.

Gómez, C. M., Pérez-Blanco, C. D., Adamson, D., and A. Loch. 2017. Managing water scarcity at a river basin scale with economic instruments. *Water Econ Policy* 4: 1750004.

Gruère, G., and H. Le Boëdec. 2019. *Navigating pathways to reform water policies in agriculture*. OECD Food, Agriculture and Fisheries Papers 128. Paris, France: OECD Publishing.

Hanak, E. 2003. *Who should be allowed to sell water in California? Third-party issues and the water market*. San Francisco: Public Policy Institute of California.

Hanemann, W. M. 2006. The economic conception of water. In *Water crisis: Myth or reality?*, ed. P. P. Rogers, M. R. Llamas, and L. Martínez-Cortina, 61–91. New York: Taylor & Francis Group.

Hassan, F. 2010. *Water history for our times*. Paris (France): UNESCO International Hydrological Programme.

Iglesias, A., Cancelliere, S., Gabiña, D., López-Francos, A., Moneo, A., and G. Rossi. 2007. *Drought management guidelines and examples of application (MEDROPLAN Project)*. Zaragoza: European Commission/MEDA Water Programme.

Intergovernmental Panel on Climate Change (IPCC). 2018. Impacts of 1.5°C of global warming on natural and human systems. In *Global warming of 1.5°C*. An IPCC special report on the impacts of global warming of 1.5°C above pre-industrial levels and related global greenhouse gas emission pathways, in the context of strengthening the global response to the threat of climate change, sustainable development, and efforts to Eradicate poverty. Geneva: IPCC.

Intergovernmental Panel on Climate Change (IPCC). 2014. *IPCC fifth assessment report (AR5)* No. WGII. Geneva, Switzerland: IPCC.

International Commission on Large Dams (ICOLD). 2018. *World register of dams*. Paris: ICOLD.

International Water Association (IWA). 2016. *Flood control and disaster management*. London, UK: IWA Publishing.

Kirhensteine, I., Cherrier, V., Jarritt, N., Farmer, A., de Paoli, G., Delacamara, G., and A. Psomas. 2016. *EU-level instruments on water reuse: Final report to support the Commission's Impact Assessment*. Luxembourg: Publications Office of the European Union.

Kraemer, R. A., Guzmán, Z., Seroa, R., and C. Russell. 2003. *Economic instruments for water management in Latin America and the Caribbean*. Executive Summary, Regional Policy Dialogue Study. Washington, DC: Inter-American Development Bank.

Lecina, S., Isidoro, D., Playán, E., and R. Aragüés. 2010. Irrigation modernization in Spain: Effects on water quantity and quality—A conceptual approach. *Int J Water Resour D* 26: 265–282.

Le Mouël, C., De Lattre-Gasquet, M., and O. Mora. 2018. *Land use and food security in 2050: A narrow road*. Versailles, France: Ed. Quæ.

Linton, J., and D. B. Brooks. 2011. Governance of transboundary aquifers: New challenges and new opportunities. *Water International* 36(5): 606–618.

Mathioulakis, E., Belessiotis, V., and E. Delyannis. 2007. Desalination by using alternative energy: Review and state-of-the-art. *Desalination* 203: 346–365.

Miller, J. E. 2003. *Review of water resources and desalination technologies*. Albuquerque, NM: Sandia National Laboratories.

Molle, F., López-Gunn, E., and van Steenbergen, F. 2018. The local and national politics of groundwater overexploitation. *Water Alternatives* 11: 445.

Molle, F., Wester, P., and P. Hirsch. 2010. River basin closure: Processes, implications and responses. *Agr Water Manage* 97: 569–577.

Mundell, R. A. 1962. The appropriate use of monetary and fiscal policy for internal and external stability. *Staff Papers* 9(1): 70–79.

Nachtergaele, F., Bruinsma, J., Valbo-Jorgensen, J., and D. Bartley. 2011. Anticipated trends in the use of global land and water resources. SOLAW Background Thematic Report–TR01. Rome: FAO.

National Center for Environmental Economics (NCEE). 2001. *The United States experience with economic incentives for protecting the environment.* Report No. EPA-240-R-01-001. Washington, DC: NCEE.

Nicoll, P. G. 2013. Forward osmosis—A brief introduction, The International Desalination Association. World Congress on Desalination and Water Reuse 2013. Tianjin, China, pp. 20–25.

Organisation for Economic Co-operation and Development (OECD). 2011. *Water governance in OECD countries: A multi-level approach.* OECD Studies on Water. Paris: OECD Publishing.

Organisation for Economic Co-operation and Development (OECD). 2012. *OECD environmental outlook to 2050: The consequences of inaction.* Paris: OECD Publishing.

Organisation for Economic Co-operation and Development (OECD). 2015a. *Water resources allocation: Sharing risks and opportunities,* OECD Studies on Water. Paris: OECD Publishing.

Organisation for Economic Co-operation and Development (OECD). 2015b. *Principles on Water Governance.* Adopted by the OECD Regional Development Policy Committee on May 11, 2015. Available at http://www.oecd.org/cfe/regional-policy/OECD-Principles-on-Water-Governance.pdf

Organisation for Economic Co-operation and Development (OECD). 2016. *OECD Fact book 2015–2016.* Paris: OECD Publishing.

Pahl-Wostl, C. 2007. Transitions towards adaptive management of water facing climate and global change. *Water Resour Manag* 21: 49–62.

Pahl-Wostl, C., Giupponi, C., Richards, K., Binder, C., de Sherbinin, A., Sprinz, D., Toonen, T., and C. van Bers. 2013. Transition towards a new global change science: Requirements for methodologies, methods, data and knowledge. *Environ Sci Policy* 28: 36–47.

Pahl-Wostl, C., Jeffrey, P., Isendahl, N., and Brugnach, M., 2010. Maturing the new water management paradigm: Progressing from aspiration to practice. *Water Resour Manag* 25: 837–856.

Perry, C., and P. Steduto. 2017. Does improved irrigation technology save water? A review of the evidence (Discussion paper on irrigation and sustainable water resources management in the Near East and North Africa). Regional Initiative on Water Scarcity for the Near East and North Africa. Cairo: FAO.

Quiggin, J., 2001. Environmental economics and the Murray–Darling river system. *Aust J Agric Resour Econ* 45: 67–94.

Randall, A. 1981. Property entitlements and pricing policies for a maturing water economy. *Aust J Agric Resour Econ* 25: 195–220.

Rey, D., Pérez-Blanco, C. D., Escriva-Bou, A., Girard, C., and T. I. E. Veldkamp. 2018. Role of economic instruments in water allocation reform: Lessons from Europe. *Int J Water Resour Dev* 35(2): 206–239.

Ryder, G. 2017. Wastewater: The untapped resource. The United Nations World Water Development Report, UN Water Programme. Paris: WWAP and UNESCO.

Santato, S., Mysiak, J., and C. D. Pérez-Blanco. 2016. The water abstraction license regime in Italy: A case for reform? *Water* 8: 1–15.

Schlosser, C. A., Strzepek, K., Gao, X., Fant, C., Blanc, E., Paltsev, S., Jacoby, H., Reilly, J., and A. Gueneau. 2014. The future of global water stress: An integrated assessment. *Earth's Future* 2(8): 341–361.

Shiklomanov, I. A., and J. C. Rodda. 2004. *World water resources at the beginning of the twenty-first century*. Cambridge: Cambridge University Press.

Sommariva, C. 2017. State of the art and future applications of desalination technologies in the Middle East. In *Water, Energy & Food Sustainability in the Middle East: The Sustainability Triangle*, ed. S. Murad, E. Baydoun, and N. Daghir, 107–124. Cham: Springer International Publishing.

Spanish Ministry of Environment. 2005. *Guía para la redacción de planes especiales de actuación en situación de alerta y eventual sequía*. Madrid: Spanish Ministry of Environment.

Spears, T. 2003. *Irrigating efficiently to feed the world in 2050*. Nebraska: Valmont Water Management Group.

Sprenger, C., Hartog, N., Hernández, M., Vilanova, E., Grützmacher, G., Scheibler, F., and S. Hannappel. 2017. Inventory of managed aquifer recharge sites in Europe: Historical development, current situation and perspectives. *Hydrogeol J* 25: 1909–1922.

Stavins, R. N. 2003. Experience with market-based environmental policy instruments. In *Handbook of environmental economics*, ed. K. G. Mäler and J. Vincent, 355–435. Amsterdam: Elsevier.

Strosser, P., Delacámara, G., Gómez, C. M., Lago, M., and A. Maziotis. 2013. Changing current practice in the application of EPIs to achieve the objectives of the WFD, EPI-Water Discussion Paper 01.

Tinbergen, J. 1952. *On the Theory of Economic Policy*. Amsterdam: North-Holland Pub. Co.

United Nations (UN). 2012. *UN-Water status report on the application of integrated approaches to water resources management for Rio+20*. New York: UN.

United Nations (UN). 2018. *World Population Prospects: The 2017 revision, key findings and advance tables*. ESA/P/WP/248. New York: UN Population Division, Department of Economic and Social Affairs.

United Nations Environment Programme (UNEP). 2009. *Integrated water resources management in action. DHI Water Policy*. Nairobi: UNEP-DHI Centre for Water and Environment.

United Nations Environment Programme (UNEP). 2016. *Transboundary Water Systems— Status and trends: Cross-cutting analysis*. Nairobi: UNEP-DHI Centre for Water and Environment.

Van Drecht, G., Bouwman, A. F., Harrison, J., and J. M. Knoop. 2009. Global nitrogen and phosphate in urban wastewater for the period 1970 to 2050. *Global Biogeochemical Cycles* 23: GB0A03.

Vollenweider, R. A. 1970. *Scientific fundamentals of the eutrophication of lakes and flowing waters, with particular reference to nitrogen and phosphorus as factors in eutrophication*. Paris: OECD Publishing.

Wada, Y., and M. F. P. Bierkens. 2014. Sustainability of global water use: Past reconstruction and future projections. *Environ Res Letters* 9(10): 104003.

Wada, Y., van Beek, L. P. H., and M. F. P. Bierkens. 2012. Nonsustainable groundwater sustaining irrigation: A global assessment. *Water Resour Res* 48(6): W00L06.

World Commission of Dams (WCD). 2000. *Dams and development. A new framework for decision-making*. London, UK: WCD.

World Energy Council (WEC). 2013. *World energy scenarios composing energy futures to 2050*. London: Project Partner Paul Scherrer Institute.

World Health Organization (WHO) and the United Nations Children's Fund (UNICEF). 2017. *Progress on drinking water, sanitation and hygiene: 2017 update and SDG baselines*. Geneva: WHO and UNICEF.

Worldwatch Institute (WWI). 2018. *Grain harvest sets record, but supplies still tight*. Washington, DC: WWI.

Yevjevich, V. 1992. Water and civilization. *Water Int* 17: 163–171.

Yoffe, S., Wolf, A., and M. Giordano. 2003. Conflict and cooperation over international freshwater resources: Indicators of basins at risk. *J Am W Resour A* 39(5): 1109–1126.

3 Underwater Vegetation, Aquatic Cultivation

*Zahra Naghibi, Jacqueline A. Stagner,
Rupp Carriveau, and David S.K. Ting*

CONTENTS

3.1 INTRODUCTION

The gap between growing populations and food supply, limited resources in terms of available water and arable land, climate change, soil degradation, urban sprawl, and a lack of distribution equity contribute to the challenge of meeting the planet's food demand [1]. Considering this increasing food demand, taking advantage of all food resources is crucial. Water ecosystems are a significant source of human food. Water, in the form of fresh water and seawater, covers more than 70% of the Earth's surface. Saving the underwater natural ecosystems plays an important role in saving the planet. Underwater vegetation or submerged aquatic vegetation (SAV), as one of the key components of aquatic ecosystems and food chains, has a crucial role in maintaining the biodiversity and overall health of ecosystems [2] and supplying food for humans.

Over recent decades, many underwater vegetation have been reported to be declining as a result of human activities [3]. It is crucial to understand the effect of human activities on aquatic vegetation to manage and protect aquatic environments. In the first part of this section, the environmental benefits of underwater vegetation, a summary of harmful activities and pollution that threatens the health of underwater plants, and some protection and restoration methods will be discussed.

Transplanting of underwater vegetation is one of the methods used for aquatic restoration. Transplanting is a type of cultivation. However, aquatic cultivation is not restricted to transplanting to restore natural ecosystems. Aquatic cultivation can be used to produce food for human consumption. Agricultural water shortages, rising sea levels as a result of global warming, and soil disappearing are some reasons for thinking about other agricultural methods. In this chapter, floating agriculture with a focus on seaweed cultivation and soilless platforms, floating greenhouses using ships, and very large floating structures (VLFSs) will be mentioned as climate change adaptation methods that are responses to the global food demand.

Another important food resource for humans is fish. Underwater vegetation has positive interactions with fishes. Many aquatic plants provide a nursery or habitat area for fish assemblages in part or all of their life cycles to protect them from predators. In addition, some underwater vegetation can serve as the food for herbivorous and omnivorous fishes. On the other hand, fish can have an important ecological role in protecting underwater vegetation. In the final part of this chapter, this positive fish-underwater vegetation interaction will be discussed.

3.2   UNDERWATER VEGETATION/UNDERWATER PLANTS

Underwater crops, as one of the main components of aquatic environments, have many environmental benefits. They can be a link in the food chain or a cover for small fish and invertebrates. In recent decades, many underwater plant species are reported to be declining worldwide [3] as a result of global warming, various kinds of pollution, and physical trauma as the consequence of shipping and scuba diving. Underwater vegetation must be protected locally and globally to maintain important ecosystem functions and structures. In this section, some environmental benefits of underwater vegetation will be discussed. Next, a summary of activities that threaten aquatic crops will be described. Finally, some methods of underwater crop protection and restoration will be discussed.

3.2.1   Environmental Benefits of Underwater Vegetation

Underwater vegetation has important environmental benefits. It can act as the base of herbivorous and detrivorous food chains [4] such as invertebrates, fish, birds, and mammals during one or more stages of their life cycle. Underwater vegetation can also be a source of organic carbon for bacteria [4,5]. In addition, some aquatic plants can provide shelter and nursery areas for prey [6]. In other words, vegetation can provide a habitat for ecological communities or ecosystems.

In addition, underwater crops can be considered the earth's carbon sink [7]. During photosynthesis, aquatic crops consume carbon dioxide (CO_2) in the water and produce oxygen. On the other hand, during respiration at night, they consume dissolved oxygen and produce CO_2. Recently, research on ponds has shown that under the optimum photosynthesis conditions, aquatic crops have potentially significant role in stabilizing the carbonate weathering-related carbon sink [7]. In the other words, there is a net CO_2 sink directly from the atmosphere to the water body. Therefore, underwater vegetation can significantly affect the terrestrial carbon budget, in some cases [7].

Storms, tsunami waves, and tropical cyclones can cause considerable damage along shorelines [8]. Aquatic crops, by buffering the wave energy and reducing flow velocity, can protect shorelines from storms and sediment erosion [9]. Wetland habitats, such as mangroves and salt marshes, can provide wave-energy and storm-surge buffers [10,11]. By reducing the bed shear stress, they are also able to trap and stabilize sediment and reduce regional erosion [12].

Some aquatic plants act as marginal filters that protect lands from the migration of contaminants and sediments. They act as a barrier for the migration of contaminants in the transition zone that connects the land and sea [13]. Contaminants could be organic pollutants, heavy metals, and pathogenic microorganisms [13]. Aquatic vegetation can also enhance water clarity by trapping sediment [14].

3.2.2 THREATS

Although underwater vegetation has many environmental benefits, over the past several decades, many aquatic vegetables in different water habitats have experienced abundant decline around the world [15]. Any reduction in aquatic vegetation can result in habitat degradation, which in turn can cause a reduction in ecosystem service functions [16]. The origin of this decline is directly or indirectly because of human activities [17]. The first key in protecting aquatic ecosystems is understanding the effect of human activities. The following is a summary of the activities that act as a threat to aquatic crops.

3.2.2.1 Chemical Pollution

Some human activities can input toxic substances into aquatic ecosystems [18]. This chemical pollution reduces the water quality or is accumulated in the sediments [19]. The sources of chemical pollution could be from outboard engines of vessels, fuel spillage, sewage, and agricultural fertilizers [20]. For example, acidification of lakes affects the accessibility of inorganic nutrients and carbon for macrophytes [21]. In addition, the concentration of some metal ions can change the underwater light availability [21].

3.2.2.2 Biological Pollution

Excessive growth of invasive alien aquatic plants (IAAPs) in marine benthic ecosystems can result in biological pollution [14]. IAAPs are considered a serious threat to global biodiversity [22] because some of these species are hazardous and compete with native plants to gain light, space, and nutrients [23]. In some cases, loss of native aquatic plants can happen as a result of invasive aquatic plants [24]. One of the main routes of this biological invasion is commercial shipping and water vessels [14]; the vessels act as a vector for invasion [25].

3.2.2.3   Physical Pollution

Physical pollution of water is caused by suspended solids and is a global danger to freshwater crops. For example, a massive influx of drifting *Sargassum*, which consists of *Sargassum fluitans* and *Sargassum natans*, in the Caribbean Sea between 2011 and 2016 was experienced [26]. These species have positive effects on ocean ecosystems; however, the beaching of them along the coastline resulted in a buildup of brown-colored, beach-cast material. Tussenbroek et al. [26] used the term "Sargassum-brown-tide" (Sbt) for this phenomena. In Figure 3.1, the effect of Sbt on the seagrass of three sites has been shown. They evaluated the effect of Sbt on light, oxygen, and pH, for four near-shore water locations.

FIGURE 3.1   Seagrass meadows before and after *Sargassum* "brown tide" at three different sites. (Reproduced with permission from the literature van Tussenbroek, B. I. et al., *Mar. Pollut. Bull.*, 122, 272–281, 2017.)

Another example of physical pollution is sediment burial. Sediment loading can cause deterioration in the water quality and underwater light, which is critical for aquatic crops [3]. In ecosystems consisting of seagrass and algae, the condition is more complicated. It has been shown that with increasing nutrient and sediment load, the seagrass system will shift toward an algae-dominated system at some point, which could be an irreversible state transition in some cases [27].

3.2.2.4   Hydrodynamic Alterations of Shipping

Shipping creates hydrodynamic alterations that threaten aquatic ecosystems. Impacts of ship-induced waves on underwater vegetation can be separated into two groups [18]: (i) impacts on the underwater vegetation growth as the result of damage and uprooting and (ii) changing species composition, distribution, and abundance. Doyle [28] experimentally studied the effect of 0.15-m waves on young *Vallisneria americana* plants. It was shown that the total mass accumulation of plants exposed to the disturbed water was 50% of the undisturbed plants. In addition, the disturbed ones had shorter leaves.

3.2.2.5   Scuba-Diving Tourism

Scuba-diving tourism has increased significantly in recent decades [29]. At the same time, the environmental interaction of scuba divers and underwater ecosystems, and environmental issues regarding this industry, have attracted considerable research interest [30,31]. Lucrezi et al. [32] summarized the negative environmental impacts of scuba diving, including direct damage to underwater ecosystems through physical contact with the habitats and indirect issues of scuba diving because of the development of coastal zones. For instance, coral reefs, destinations for many recreational divers, have experienced a significant decline over recent decades as a result of physical contact from divers [33].

3.2.2.6   Marine Structures

The increasing number of marine and coastal structures in water environments is causing "ocean sprawl" [34]. Ocean sprawl can have a crucial impact on natural ecosystems. Structures built over underwater vegetation may shade and make a physical barrier for the light and, as a result, kill the macrophytes [34]. These impacts are not only restricted to the location of the structures but can also have a critical impact on the regional scale [11]. As an example, construction of low-crested structures to defend shores from erosion have some local impacts, such as disruption of surrounding soft-bottom environments, and regional impacts on regional species diversity and favoring the spread of non-native species [35].

3.2.2.7   Global Warming

As a result of global warming, world water temperatures have increased significantly in recent decades. According to the Intergovernmental Panel on Climate Change (IPCC), sea surface temperatures have also seen a rise of 0.4°C–0.8°C in the past century [36]. Water temperature rise can have impacts on individual, plant-herbivore interactions, or native plant-invasive aquatic plant interactions.

In recent decades, many species of underwater vegetation have experienced a rapid decline as a result of increased water temperatures. For example, coral reef cover has experienced a rapid decline due to increased water temperatures across the Caribbean region [37]. In addition, rising water temperatures influence the interactions between species. Pages et al. [38] showed that as the result of changing the complex factors determining species interaction in the warming Mediterranean, there will be a clear set of winners and losers. The response of selected plant and animal interactions are shown in Figure 3.2. Their proposed model can evaluate the effect of warming on the interaction of the plants (*Posidonia oceanica* seagrass, seagrass *Cymodocea nodosa*, and macroalga *Cystoseira mediterranea*) and herbivores (sea urchins).

To investigate the effect of global warming on native plants and invasive alien plants, Verlinden et al. [39] considered two pairs of *Senecio inaequidens—Plantago lanceolata* (alien invasive—native) with the domination of the alien species and *Solidago gigantea—Epilobium hirsutum* with the domination of the native species. The results showed that rising temperatures caused invasion-dominance reduction in the first pair, and the dominance of the invader increased in the second pair. According to their study, the effect of warming on invasive plants must be considered with their interaction with native plants. Monoculture behavior is not a sufficient indicator of plant response to climate change.

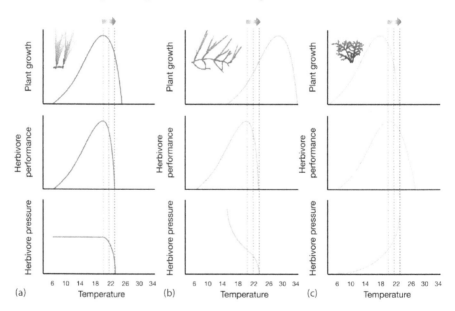

FIGURE 3.2 The outcomes of plant-herbivore interactions in a warming Mediterranean. Arrows show the direction of warming. (a) The herbivore pressure on *Posidonia oceanica* seagrass remained unchanged with warming; however, the decrease in herbivore happened when warming increased because with rising temperature, the *P. Oceanica* seagrass leaves incubated. (b) The herbivore pressure decreased for seagrass *Cymodocea nodosa* during warming. (c) The herbivore pressure increased for macroalga *Cystoseira mediterranea* during warming because of the low performance of the macroalga at warm temperatures. (Reproduced with permission from the literature Pagès, J.F. et al., *Mar. Pollut. Bull.*, 134, 55–65, 2018.)

3.2.3 UNDERWATER CROP PROTECTION AND RESTORATION

The global decline in the underwater vegetation demands urgent management strategies to protect them. Consideration of marine-protected areas (MPAs), land-use patterns, and transplantation are some important methods of underwater crop protection or restoration that will be discussed in this section. These methods can prevent the harmful human activities motioned in the previous section and have an important role in the restoration of the threatened area.

The creation of MPAs is an important management tool to reduce the trend of biodiversity loss in marine ecosystems [40]. The International Union for Conservation of Nature (IUCN) defined the term "marine protected area" as "any area of intertidal or subtidal terrain, together with its overlying waters and associated flora, fauna, historical and cultural features, which has been reserved by legislation to protect part or all of the enclosed environment" [41]. According to the IUCN, these areas should protect against outside activities, and depleted, threatened, rare, or endangered species should be protected in these areas [41]. According to the United Nations Environment Programme (UNEP), there has been considerable growth in MPAs across the world in recent years [42]. The current percentages of MPAs are shown in Figure 3.3. In July 2018, 45.7% of nearshore marine ecoregions have been met the marine protection target of 10%.

Quiros et al. [43] showed that land use can be more important than marine protection in some cases. In their study, they considered the relationship between terrestrial and marine systems to evaluate the effects of human land-based activities on seagrasses. They found that human activities on land have major impacts on the abundance and richness of seagrass. They saw this pattern across a large spatial scale (more than 800 km) and more than 54 samples of seagrass meadow.

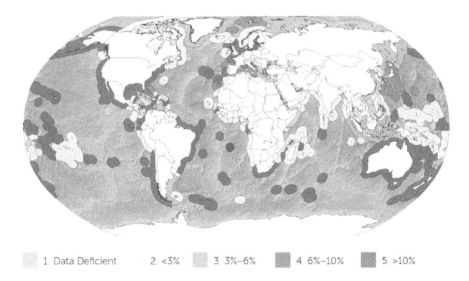

1. Data Deficient   2. <3%   3. 3%–6%   4. 6%–10%   5. >10%

FIGURE 3.3 Percentage of protected marine areas coverage. (Reproduced with permission from the literature UNEP-WCMC, IUCN, and NGS, "Protected Planet Report 2018," UNEP-WCMC, IUCN and NGS: Cambridge, UK; Gland, Switzerland; and Washington, DC, USA, 2018.)

The management of the integration of terrestrial and aquatic environments has been proven in other studies, as well. Employment of concurrent aquatic and terrestrial conservation actions is delivering cost-effective outcomes for marine conservation in comparison with other land- or sea-based conservation actions to protect aquatic ecosystems [44].

In addition to the management and conservation of aquatic vegetables, restoration actions are needed in marked habitat areas with degradation and reduction of aquatic vegetation across the world [16]. Transplantation of some aquatic vegetation is recognized as the best method for their restoration. For example, it is a conventional method in the restoration of coral reefs. Jaap (2000) [45] summarized steps for the transplantation of physically damaged coral reefs. These steps include eliminating loose debris from the reef, rebuilding three-dimensional structures into leveled-scarified reef surfaces, and transplanting reefs back. However, some deficiencies have been mentioned for the transplanting method. Four of them were mentioned by Li et al. (2008) [46]. First, low efficiency and being labor-intensive limit its application for large scales. Second, this method usually causes great harm to the underwater vegetation at the source site. Third, transplanting can be considered a "legal" way of spreading invasive aquatic plants, and because of that, in some cases finding native macrophytes has become more difficult. Fourth, the plants restored with the transplanting are not stable.

Monitoring the impacts of human activities on underwater health conditions helps to understand the interaction of human and underwater vegetation and improve the management restoration and protection tools by the time. In this way, education also plays an important role in underwater crop protection. The more humans learn about the underwater ecosystem threats and ways of protection, the better prepared they will be to make deep and lasting changes. All methods of protection and restoration have their advantages and disadvantages, and the selection of the best method depends on the type of crop which needs to be restored. Depending on the environmental issue that aquatic plants are struggling with and the type of underwater vegetation, different approaches or methods will be adopted to restore the aquatic plants.

3.3   AQUATIC CULTIVATION OF CROPS

In the previous section, the transplanting of aquatic plants was mentioned as a method for the restoration of aquatic ecosystems. In addition to the restoration of aquatic ecosystems, aquatic cultivation can be employed to produce plants to be used in various applications such as food, pet food, fertilizers, and cosmetics [47]. It has been proven that some edible sea vegetables are significantly richer in nutrients compared to terrestrial plants [48]. Water scarcity is another motivation to pay attention to aquatic agriculture. In recent decades, agricultural water shortages have been becoming serious problems in many locations of the world [49]. As agriculture is a water-demanding part of human activities, water scarcity is jeopardizing irrigated agriculture [50]. Considering this condition, using water-based environments as a potential solution to water shortages in the agricultural sector has been proposed [51].

Aquatic cultivation could be a climate change adaption in flood-prone areas. Chowdhury et al. [52] considered global climate change as another reason for consideration of aquatic cultivation. According to their study, floods and waterlogging, as a result of global climate change, are threatening Bangladesh, a flood-prone country. The agricultural sector is extremely damaged in these areas. In their study area, long-term waterlogging makes traditional, land-based agriculture difficult. Floating farms, as a "self-innovated farming technique," have been employed by local farmers to counter this problem. This method enables local farmers to retain their lives and livelihoods in flooded areas [52]. Floating agriculture will be explained in the next section.

Therefore, as a result of global climate change and its consequences, including water levels rising, soil disappearing, and water shortages, other methods of nontraditional agriculture should be considered. Aquatic agriculture will be explained in the following sections. First, floating agriculture will be introduced, with focus on seaweed cultivation and soilless floating platforms. Next, floating greenhouses, using ships and VLFS, will be mentioned.

3.3.1 FLOATING AGRICULTURE

3.3.1.1 Seaweed Cultivation

Seaweed is another term for marine microalgae. Algae are plant organisms because they have chlorophyll that plays a role in photosynthesis for the production of the organic matter and oxygen in waters [53]. However, there is some debate as to whether algae should be considered plants. Some researchers believe that algae are not plants and call them *protist* or *protoctist* [54]. They are rich sources of nutrients in human food. However, their applications are not limited to the food industry. Algae can be classified into two groups: unicellular organisms (microalgae) and multicellular organisms (macroalgae) [48]. Seaweeds are classified into four separate groups based on their color: red algae, brown algae, green algae, and blue-green algae.

Among all methods of seaweed farming, two have been recognized as floating agriculture: floating lines and floating rafts. According to de San's study [55], in floating-line (long-line) systems, lines with a maximum length of 50 m are used. They are anchored at each end and have a float attached approximately every 10 m to support the line. Robledo et al. [56] described a floating-raft system with a 10- $\times$ 20-m module built with a net of bamboo poles (at intervals of approximately 5 m) and polypropylene ropes (at intervals of approximately 1 m). The details of the system are illustrated in Figure 3.4. Polypropylene ropes are used as the cultivation lines. For anchoring to the seabed, 50-kg weights are used. To keep the cultivation module at a depth of 25–30 cm under the water surface as the biomass grows, additional floatation buoys are added to the system. For seeding, maintenance, and harvesting, a boat without any engine can be used.

3.3.1.2 Floating Gardening

Floating gardening is a form of hydroponics [57]. Hydroponics is a method of agriculture in which plants grow without using soil. In this method, plants use mineral nutrient solutions in water instead of soil [52,58]. These nutrients are based on potassium (K+) and nitrate (NO_{3-}) ions [59]. The bed, which is used for keeping the

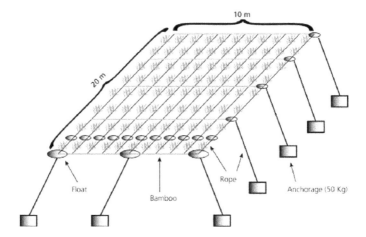

FIGURE 3.4   Seaweed cultivation using a floating raft system. (From Food and Agriculture Organization of the United Nations. Reproduced with permission from the literature Robledo, D., *Inst. Politécnico Nac.*, 2013.)

plant inert, could be made of various materials such as gravel, sand, perlite, etc. [52]. When the bed is floated on the water, it is called *floating gardening*. Depending on the region, the raw material that is used for construction of the floating bed can vary [52].

As mentioned in previous sections, this method can be considered as a climate change adaptation method in flood-prone regions. Chowdhury et al. [52] reviewed the possibility of using this agricultural technique as a climate change adaptation method in Bangladesh. This technique is an ancient method of crop cultivation in southern parts of Bangladesh [57]. In Figure 3.5, the application of using water hyacinth (*Eichhornia crassipes*) as the floating bed material is shown [60], and in most regions of Bangladesh, water hyacinth is used as the main part of the floating platforms [52]. In other regions, using aquatic weeds, paddy straw, and coconut fiber is quite common [52]. The study by Chowdhury et al. [52] showed that floating agriculture can support the local farmers to maintain their lives and livelihoods during some environmental disasters such as floods and long-term waterlogged conditions. They found that floating agriculture is not only effective as an agricultural method but is also environmentally sound, economically possible, and socially feasible.

3.3.2   FLOATING GREENHOUSE/FLOATING FARM

Floating greenhouses can be considered one form of urban agriculture [61]. Floating greenhouses are a type of VLFS or very large floating platform (VLFP). VLFSs are defined as "artificially man-made floating land parcels on the sea" [62] that can be spread in coastal or offshore regions [63]. The various applications of VLSFs have attracted the attention of many researchers in recent decades [62].

According to Van der Pol's study [64], employing VLFSs or "building with water" is a "flood-proof architecture." Actually, the novel concept of "living with

FIGURE 3.5 Using water hyacinth and other aquatic weeds as material for a floating bed. (From Community-led adaptation in Bangladesh. Reproduced with permission from the literature Pender, J., *Forced Migr. Rev.*, 31, 54–55, 2008.)

water" is a kind of climate-change adaptation and could help communities to respond to rising water levels. VLFSs are helpful in creating space on the sea for infrastructure and building for coastal cities with scarce land [65]. Floating greenhouses can offer the opportunity of combining greenhouse horticulture and water storage in the same place that could benefit space-restricted areas [64]. From a structural point of view, because VLFSs are base-isolated structures, they are resistant to seismic impact [66].

The combination of ease of renewable energy production and very-low-energy demand in VLFSs is another of their advantages [67]. Solar-thermal and photovoltaic systems, passive solar design, wind turbines, and seawater heat pumps have been implemented on VLFSs [67]. In some cases, floating structures can freely turn toward the sun to collect the most solar energy [68]. As an example, the Science Barge, a prototype of a sustainable floating greenhouse, constructed by the New York Sun Works Center at the Hudson River in Manhattan is shown in Figure 3.6. It is a demonstration of urban agriculture on a flouting structure. Products of this greenhouse are tomatoes, melons, corn, peppers, eggplants, and lettuce. The required energy is provided from solar energy, wind energy, and biofuels. The irrigation water is provided by collected rainwater and purified river water.

If the floating greenhouse is on seawater, desalination is necessary in the majority of cases. Moustafa [69] proposed a combination of farm- and solar-energy–based seawater desalination in the same floating platform. This system was called "bluehouse." The conceptual design of his proposed system is shown in Figure 3.7. This system has three main parts: (i) a sunlight-harvesting unit (or photovoltaic system) that absorbs sunlight and converts it into electricity, (ii) a thermal desalination unit to desalinate

FIGURE 3.6 Using renewable energy resources in Science Barge, experimental floating greenhouses. (Reproduced with permission from the literature Wang, C.M. and Tay, Z.Y., *Procedia Eng.*, 14, 62–72, 2011.)

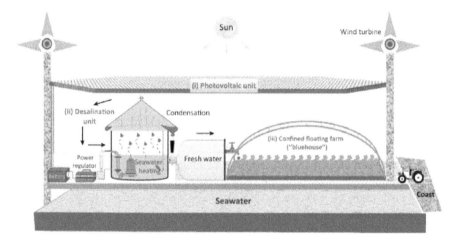

FIGURE 3.7 The proposed hybrid agricultural system by Moustafa. (Reproduced with permission from the literature Moustafa, K., *Trends Biotechnol.*, 34, 257–259, 2016.)

seawater using the energy produced by the photovoltaic system, and (iii) a "floating farm" unit with an arable surface. In addition to the ability to adapt to global climate change, the isolated floating farms would reduce the impact of plant diseases and the need for fertilizers. However, a disadvantage of such systems is their high cost, especially in the initial phases of their development. Another drawback is that this system is suitable only in dry, sunny regions, which is based on a photovoltaic.

Another floating farm with seawater desalination system has been proposed by Wang et al. [51]. They introduced the idea of combining the floating film and solar desalination systems. The outline of their proposed system is illustrated in Figure 3.8. The internal surface of the film concentrates the sunlight and achieves the heat that is necessary to evaporate seawater and complete the desalination process. The purified water can be used for irrigation purposes.

FIGURE 3.8 Scheme of floating film with solar desalination system. (Reproduced with permission from the literature Wang, Q. et al., *Appl. Energy*, 224, 510–526, 2018.)

The stability of these floating surfaces is a crucial consideration. Wang and Tay [62] reviewed the VLFS technology. They categorized the VLFSs into two major types: semisubmersible type and pontoon type. The semisubmersible type has column tubes that help the structure to remain above sea level. This type is usually used in deep water with large waves. The pontoon type can remain on the water surface and is suitable for calm water. According to their study, the pontoon type behaves elastically under wave action because of its large surface area and relatively small depth. To make sure that the floating structure is stable in its position, an appropriate mooring system has to be designed. Figure 3.9 illustrates different types of the mooring system. Two main types of mooring systems in floating structures were introduced: the mooring line type and the caisson or pile-type.

Before the addition of a large artificial floating structure to the marine environment, many environmental assessments must be considered. Local losses, such as degradation of natural ecosystems and their associated species, can be some significant environmental impacts of VLFS [11]. The environmental impacts of these marine structures were discussed in previous sections. These impacts are not limited to the construction sites only; environmental effects can propagate on a large scale because of the ecological connectivity and genetic structure of populations [11]. Therefore, to effectively design a floating farm and reduce its ecological footprint, extensive investigations are necessary. These investigations must be in marine engineering, economics, biotechnology, ecosystem biology, etc. [69]. In this kind of design, an

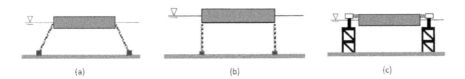

(a) (b) (c)

FIGURE 3.9 Different types of mooring systems of very large floating structures (VLFSs): (a) Cable/Chain as the mooring line type, (b) Tension Leg as the mooring line type, and (c) Rubber Fender-Dolphin as the caisson or pile type. (Reproduced with permission from the literature Wang, C.M. and Tay, Z.Y., *Procedia Eng.*, 14, 62–72, 2011.)

eco-engineering framework is needed to minimize the ecological impact of structures [11]. In other words, a balance between engineering requirements and ecological ecosystem needs is necessary.

3.4   UNDERWATER VEGETATION AND FISHERY

Because fish are the main source of animal protein for about 1.25 thousand billion people within 39 countries worldwide, they play a significant role in world food supply [1]. Using fish in food production is a response to rising food demand [70]. There is a positive interaction between fish and the associated underwater vegetation. Aquatic vegetation can provide shelter and food for herbivorous fish. More than 60% of fish for human consumption is from herbivorous and omnivorous species [71]. These fish can help to maintain a balance between underwater plants by avoiding seaweed overgrowth and outbreaks. This interaction will be discussed more in this section.

Various underwater-vegetated habitats are used by fish assemblages in part or all of their life cycle to provide shelter or food. For example, coral reefs serve as habitat and nursery areas for many juvenile fishes and the fishing industry highly depends on coral reefs [72,73]. Although coral reefs cover only 0.1%–0.5% of the ocean floor, almost one-third of marine fish species are found on coral reefs [74], and fish living in the reef areas constitute 10% of fish consumed by humans [74]. In addition, algae are one of the main food items of marine herbivorous fish [75]. Some macroalgae have been mentioned to be refuges and feeding habitats for fish [75–79]. The macroalgae *Oedogonium* can meet the nutritional requirements of herbivorous fish as a result of its high energy potential, high lipid concentrate, high essential amino acid levels, and high protein concentration [71]. Kelp can provide food both directly using its tissues and indirectly using associated understory algae for herbivorous species [78]. For instance, *Ayu Plecoglossus altivelis* is a herbivorous fish species that lives throughout the Japanese archipelago [80]. This fish, after completing the marine part of its life cycle during its juvenile stage in winter, migrates into rivers to graze on benthic algae attached to the riverbed [80]. *Ayu* is commonly targeted for commercial fishing and is a significant component of the food chain in many Japanese streams [80].

Several commercial and recreational fishes choose seagrass beds as feeding and nursery areas because of their highly complex structure [74,81]. Seagrass habitats, by providing vegetative habitats and shelter for fish, are highly productive areas [81]. Several studies have compared the utilization of seagrass habitats and unvegetated habitats on the abundance and diversity of fish assemblages. Park and Kwak [81] studied the influences of seagrass vegetation on fish assemblage structures by collecting 54 fish species belonging to 30 families at both seagrass beds and unvegetated habitats on the coast of Namhae Island, South Korea. They concluded that habitat type and season highly affected the number of species, their abundance, and the diversity of fish assemblages. Fish abundance peaked in seagrass habitats during the spring season and was lowest in unvegetated areas during the winter season. Bloomfield and Gillanders [82] showed that, among various habitats, seagrass supports the highest numbers of fish and the greatest species richness. Therefore, fish assemblage structures are significantly affected by habitat type [81] and saving the natural habitat of particular species that are dependent on specific habitats for part or all of their life cycle is crucial to saving the fish species richness and abundance [82].

Herbivorous fish play a major ecological role in the protection of underwater vegetation such as macroalgae and coral reefs. Ruz et al. [78] focus on positive feedbacks that a herbivorous fish, *Aplodactylus punctatus*, may provide to benefit a special kelp, *Lessonia trabeculata*. According to their study, the herbivorous fish may have the potential to be a facilitator of kelp zoospore transport, and it can be effective in natural kelp population recovery. Moreover, herbivorous fish can help to maintain a balance between corals and macroalgae on reefs. Macroalgae are extremely fast growing and less sensitive to pollution and climate change than coral. Therefore, macroalgae tend to be dominant in these ecosystems. However, herbivorous fish, by grazing on the macroalgae, are helpful in maintaining a healthy competition between them, and they prevent macroalgae overgrowth and ecological shifts [83].

Overall, in addition to the fact that edible underwater vegetation can be used as food, paying attention to the health of underwater vegetation is important to save fish, as another abundant and nutritious form of food. As has been mentioned, fish highly depend on underwater vegetation for food and habitat and underwater vegetation depends on the fish grazing for reproductive success and helping the dispersal mechanism. Therefore, paying attention to maintaining a productive underwater environment is necessary to guarantee the continued existence of fish as a food source.

3.5 CONCLUSION

To achieve wider global-food security among all environmental problems that humanity faces, focusing on water-based food resources seems crucial. Underwater vegetation acts as the base of herbivorous and detrivorous food chains. Thus, protection and restoration of them have a significant role in securing the underwater resources. Introduction of MPAs, land-use patterns, and transplantation are important methods to reduce the trend of biodiversity loss in marine ecosystems. Transplanting of aquatic plants and underwater cultivation are not only for the restoration of the aquatic system but can also be employed to produce plants with the main application in the food industry. Moreover, it can be a response to problems that climate change and water scarcity made in the agricultural sector. The ease of renewable energy harvesting and its very low energy demand are another advantages of aquatic cultivation. However, to widely employ these methods of food production, extensive environmental assessments are required.

ACKNOWLEDGMENT

We acknowledge the support of the Natural Sciences and Engineering Research Council of Canada (NSERC).

REFERENCES

1. K. Ma, S. Khan, and K. Miyan, "Aquaculture as a food production system: A review," *Biol. Med.*, vol. 3, no. 2, pp. 291–302, 2011.
2. G. A. Geevarghese, B. Akhil, G. Magesh, P. Krishnan, R. Purvaja, and R. Ramesh, "A comprehensive geospatial assessment of seagrass distribution in India," *Ocean Coast. Manag.*, no. 159, pp. 16–25, 2018.
3. Y. Zhang, X. Liu, B. Qin, K. Shi, J. Deng, and Y. Zhou, "Aquatic vegetation in response to increased eutrophication and degraded light climate in Eastern Lake Taihu: Implications for lake ecological restoration," *Sci. Rep.*, vol. 6, p. 23867, 2016.
4. E. Rejmankova, "The role of macrophytes in wetland ecosystems," *J. Ecol. F. Biol.*, vol. 34, no. 4, pp. 333–345, 2011.
5. A. R. Hughes, S. L. Williams, C. M. Duarte, K. L. Heck, and M. Waycott, "Associations of concern: Declining seagrasses and threatened dependent species," *Front. Ecol. Environ.*, vol. 7, no. 5, pp. 242–246, 2009.
6. T. Leslie, N. C. James, W. M. Potts, and A. Rajkaran, "The relationship between habitat complexity and nursery provision for an estuarine-dependent fish species in a permanently open South African Estuary," *Estuar. Coast. Shelf Sci.*, vol. 198, pp. 183–192, 2017.
7. B. Chen et al., "Coupled control of land uses and aquatic biological processes on the diurnal hydrochemical variations in the five ponds at the Shawan Karst Test Site, China: Implications for the carbonate weathering-related carbon sink," *Chem. Geol.*, vol. 456, pp. 58–71, 2017.
8. K. B. Gedan, M. L. Kirwan, E. Wolanski, E. B. Barbier, and B. R. Silliman, "The present and future role of coastal wetland vegetation in protecting shorelines: Answering recent challenges to the paradigm," *Clim. Change*, vol. 106, no. 1, pp. 7–29, 2011.
9. M. Paul, "The protection of sandy shores—Can we afford to ignore the contribution of seagrass?," *Mar. Pollut. Bull.*, vol. 134, no. May, pp. 152–159, 2017.
10. N. H. Tri, W. N. Adger, and P. M. Kelly, "Natural resource management in mitigating climate impacts: The example of mangrove restoration in Vietnam," *Glob. Environ. Chang.*, vol. 8, no. 1, pp. 49–61, 1998.
11. M. Mayer-Pinto et al., "Building 'blue': An eco-engineering framework for foreshore developments," *J. Environ. Manage.*, vol. 189, pp. 109–114, 2017.
12. M. S. Fonseca and J. S. Fisher, "A comparison of canopy friction and sediment movement between four species of seagrass with reference to their ecology and restoration," *Mar. Ecol. Prog. Ser.*, vol. 29, no. 1, pp. 15–22, 1986.
13. K. Obolewski and M. Bąkowska, "Epiphytic invertebrate patterns in coastal lakes along a gradient of salinity and water exchange with the sea," *Estuar. Coast. Shelf Sci.*, vol. 197, pp. 150–158, 2017.
14. A. Hussner et al., "Management and control methods of invasive alien freshwater aquatic plants: A review," *Aquat. Bot.*, vol. 136, pp. 112–137, 2017.
15. D. Hesley, D. Burdeno, C. Drury, S. Schopmeyer, and D. Lirman, "Citizen science benefits coral reef restoration activities," *J. Nat. Conserv.*, vol. 40, pp. 94–99, 2017.
16. Y. Zhang et al., "Global loss of aquatic vegetation in lakes," *Earth-Science Rev.*, vol. 173, pp. 259–265, 2017.
17. M. Waycott et al., "Accelerating loss of seagrasses across the globe threatens coastal ecosystems," *Proc. Natl. Acad. Sci.*, vol. 106, no. 30, pp. 12377–12381, 2009.
18. F. Gabel, S. Lorenz, and S. Stoll, "Effects of ship-induced waves on aquatic ecosystems," *Sci. Total Environ.*, vol. 601, pp. 926–939, 2017.
19. S. Burgin and N. Hardiman, "The direct physical, chemical and biotic impacts on Australian coastal waters due to recreational boating," *Biodivers. Conserv.*, vol. 20, no. 4, pp. 683–701, 2011.

20. M. J. Liddle and H. R. A. Scorgie, "The effects of recreation on freshwater plants and animals: A review," *Biol. Conserv.*, vol. 17, no. 3, pp. 183–206, 1980.

21. A. M. Farmer, "The effects of lake acidification on aquatic macrophytes: A review," *Environ. Pollut.*, vol. 65, no. 3, pp. 219–240, 1990.

22. O. E. Sala et al., "Global Biodiversity Scenarios for the Year 2100," *Science*, vol. 287, no. 5459, pp. 1770–1774, 2000.

23. N. Cantasano, G. Pellicone, and V. Di Martino, "The spread of *Caulerpa cylindracea* in Calabria (Italy) and the effects of shipping activities," *Ocean Coast. Manag.*, vol. 144, pp. 51–58, 2017.

24. I. Stiers, N. Crohain, G. Josens, and L. Triest, "Impact of three aquatic invasive species on native plants and macroinvertebrates in temperate ponds," *Biol. Invasions*, vol. 13, no. 12, pp. 2715–2726, 2011.

25. D. Früh, S. Stoll, and P. Haase, "Physicochemical and morphological degradation of stream and river habitats increases invasion risk," *Biol. Invasions*, vol. 14, no. 11, pp. 2243–2253, 2012.

26. B. I. van Tussenbroek et al., "Severe impacts of brown tides caused by *Sargassum* spp. on near-shore Caribbean seagrass communities," *Mar. Pollut. Bull.*, vol. 122, no. 1–2, pp. 272–281, 2017.

27. M. M. van Katwijk et al., "Early warning indicators for river nutrient and sediment loads in tropical seagrass beds: A benchmark from a near-pristine archipelago in Indonesia," *Mar. Pollut. Bull.*, vol. 62, no. 7, pp. 1512–1520, 2011.

28. R. D. Doyle, "Effects of waves on the early growth of *Vallisneria americana*," *Freshw. Biol.*, vol. 46, no. 3, pp. 389–397, 2001.

29. J. de Groot and S. R. Bush, "The potential for dive tourism led entrepreneurial marine protected areas in Curacao," *Mar. Policy*, vol. 34, no. 5, pp. 1051–1059, 2010.

30. S. Lucrezi et al., "Scuba diving tourism systems and sustainability: Perceptions by the scuba diving industry in two Marine Protected Areas," *Tour. Manag.*, vol. 59, pp. 385–403, 2017.

31. K. Dimmock and G. Musa, "Scuba Diving Tourism System: A framework for collaborative management and sustainability," *Mar. Policy*, vol. 54, pp. 52–58, 2015.

32. S. Lucrezi and M. Saayman, "Sustainable scuba diving tourism and resource use: Perspectives and experiences of operators in Mozambique and Italy," *J. Clean. Prod.*, vol. 168, pp. 632–644, 2017.

33. J. Toyoshima and K. Nadaoka, "Importance of environmental briefing and buoyancy control on reducing negative impacts of SCUBA diving on coral reefs," *Ocean Coast. Manag.*, vol. 116, pp. 20–26, 2015.

34. M. J. Bishop et al., "Effects of ocean sprawl on ecological connectivity: Impacts and solutions," *J. Exp. Mar. Bio. Ecol.*, vol. 492, pp. 7–30, 2017.

35. L. Airoldi et al., "An ecological perspective on the deployment and design of low-crested and other hard coastal defence structures," *Coast. Eng.*, vol. 52, no. 10–11, pp. 1073–1087, 2005.

36. J. Hallett, "Climate change 2001: The scientific basis. Edited by J. T. Houghton, Y. Ding, D. J. Griggs, N. Noguer, P. J. van der Linden, D. Xiaosu, K. Maskell and C. A. Johnson. Contribution of Working Group I to the Third Assessment Report of the Intergovernmental Panel on Climate change," *Q. J. R. Meteorol. Soc. A J. Atmos. Sci. Appl. Meteorol. Phys. Oceanogr.* vol. 128, no. 581, pp. 1038–1039, 2002.

37. S. J. Pittman, M. Poti, C. F. G. Jeffrey, L. M. Kracker, and A. Mabrouk, "Decision support framework for the prioritization of coral reefs in the U.S. Virgin Islands," *Ecol. Inform.*, vol. 47, pp. 26–34, 2018.

38. J. F. Pagès et al., "Contrasting effects of ocean warming on different components of plant-herbivore interactions," *Mar. Pollut. Bull.*, vol. 134, pp. 55–65, 2018.

39. M. Verlinden, H. J. De Boeck, and I. Nijs, "Climate warming alters competition between two highly invasive alien plant species and dominant native competitors," *Weed Res.*, vol. 54, no. 3, pp. 234–244, 2014.

40. L. J. Wood, L. Fish, J. Laughren, and D. Pauly, "Assessing progress towards global marine protection targets: Shortfalls in information and action," *Oryx*, vol. 42, no. 3, pp. 340–351, 2008.

41. Đ. Kelleher, *Guidelines for marine protected areas*, IUCN: Gland, Switzerland and Cambridge, UK, 1999.

42. UNEP-WCMC, IUCN, and NGS, "Protected Planet Report 2018," UNEP-WCMC, IUCN and NGS: Cambridge, UK; Gland, Switzerland; and Washington, DC, 2018.

43. T. E. A. L. Quiros, D. Croll, B. Tershy, M. D. Fortes, and P. Raimondi, "Land use is a better predictor of tropical seagrass condition than marine protection," *Biol. Conserv.*, vol. 209, pp. 454–463, 2017.

44. C. J. Klein et al., "Prioritizing land and sea conservation investments to protect coral reefs," *PLoS One*, vol. 5, no. 8, p. e12431, 2010.

45. J. Walter C., "Coral Reef Restoration Handbook," *Ecol. Eng.*, vol. 15, pp. 345–364, 2000.

46. E. H. Li, G. H. Liu, W. Li, L. Y. Yuan, and S. C. Li, "The seed-bank of a lakeshore wetland in Lake Honghu: Implications for restoration," *Plant Ecol.*, vol. 195, no. 1, pp. 69–76, 2008.

47. Ý. J. McHugh, "A guide to the seaweed industry, FAO Fisheries Technical Paper 441," FAO: Rome, 2003.

48. Đ. Gallo, "Novel foods: Algae," in P. Ferranti, E. M. Berry, and J. R. Anderson (eds.) *Encyclopedia of Food Security and Sustainability*. Elsevier: Amsterdam, Netherlands; Kidlington, Oxford; and Cambridge, MA, 2019.

49. A. D. Leroux, V. L. Martin, and H. Zheng, "Addressing water shortages by force of habit," *Resour. Energy Econ.*, vol. 53, pp. 42–61, 2018.

50. V. Martínez-Alvarez, M. J. González-Ortega, B. Martin-Gorriz, M. Soto-García, and J. F. Maestre-Valero, "The use of desalinated seawater for crop irrigation in the Segura River Basin (south-eastern Spain)," *Desalination*, vol. 422, pp. 153–164, 2017.

51. Q. Wang, Z. Zhu, G. Wu, X. Zhang, and H. Zheng, "Energy analysis and experimental verification of a solar freshwater self-produced ecological film floating on the sea," *Appl. Energy*, vol. 224, pp. 510–526, 2018.

52. R. B. Chowdhury and G. A. Moore, "Floating agriculture: A potential cleaner production technique for climate change adaptation and sustainable community development in Bangladesh," *J. Clean. Prod.*, vol. 150, pp. 371–389, 2017.

53. N. Andriyani, A. Mahdiana, R. Dewi, Y. Subagyo, and T. Junaidi, "Biodiversity of algae potentially HABS (Harmful Algae Blooms) in reservoir Mrica, Banjarnegara," in *E3S Web of Conferences*, vol. 47, p. 03001, 2018.

54. M. D. Guiry, "What are algae?," 2018. Available: http://www.seaweed.ie/algae/algae. php. [Accessed: November 22, 2018].

55. M. De San, "The farming of seaweeds," Smart Fish Programme Report SF/2012/30, 2012.

56. D. Valderrama, J. Cai, N. Hishamunda, and N. Ridler (eds.) *Social and economic dimensions of carrageenan seaweed farming*, FAO Fisheries and Aquaculture Technical Paper 580. FAO: Rome, 2013.

57. H. M. Irfanullah, A. Adrika, A. Ghani, Z. A. Khan, and M. A. Rashid, "Introduction of floating gardening in the north-eastern wetlands of Bangladesh for nutritional security and sustainable livelihood," *Renew. Agric. Food Syst.*, vol. 23, no. 2, pp. 89–96, 2008.

58. A. M. Pringle, R. M. Handler, and J. M. Pearce, "Aquavoltaics: Synergies for dual use of water area for solar photovoltaic electricity generation and aquaculture," *Renew. Sustain. Energy Rev.*, vol. 80, pp. 572–584, 2017.

59. G. Dini, E. Princi, S. Gamberini, and L. Gamberini, "Nemo's Garden: Growing plants underwater," *Ocean. 2016 MTS/IEEE Monterey, IEEE 2016*, pp. 1–6, 2016.
60. J. Pender, "Community-led adaptation in Bangladesh," *Forced Migr. Rev.*, vol. 31, pp. 54–55, 2008.
61. R. Aerts, V. Dewaelheyns, and W. M. J. Achten, "Potential ecosystem services of urban agriculture: A review," *PeerJ Prepr.*, no. e2286v1, 2016.
62. C. M. Wang and Z. Y. Tay, "Very large floating structures: Applications, research and development," *Procedia Eng.*, vol. 14, pp. 62–72, 2011.
63. M. Lamas-Pardo, G. Iglesias, and L. Carral, "A review of Very Large Floating Structures (VLFS) for coastal and offshore uses," *Ocean Eng.*, vol. 109, pp. 677–690, 2015.
64. J. Van der Pol, "Flood proof architecture," *Climate of Coastal Cooperation. Netherlands: Coastal and Marine Union*, 2011.
65. H. P. Nguyen, J. Dai, C. M. Wang, K. K. Ang, and V. H. Luong, "Reducing hydroelastic responses of pontoon-type VLFS using vertical elastic mooring lines," *Mar. Struct.*, vol. 59, pp. 251–270, 2018.
66. M. Riyansyah, C. M. Wang, and Y. S. Choo, "Connection design for two-floating beam system for minimum hydroelastic response," *Mar. Struct.*, vol. 23, no. 1, pp. 67–87, 2010.
67. A. Dulic, J. Angel, and S. Sheppard, "Designing futures: Inquiry in climate change communication," *Futures*, vol. 81, pp. 54–67, 2016.
68. E. E. M. Luiten et al., "Realizing the promises of marine biotechnology," *Biomol. Eng.*, vol. 20, no. 4–6, pp. 429–439, 2003.
69. K. Moustafa, "Toward future photovoltaic-based agriculture in sea," *Trends Biotechnol.*, vol. 34, no. 4, pp. 257–259, 2016.
70. J. P. Fry et al., "Environmental health impacts of feeding crops to farmed fish," *Environ. Int.*, vol. 91, pp. 201–214, 2016.
71. M. J. Vucko, A. J. Cole, J. A. Moorhead, J. Pit, and R. De Nys, "The freshwater macroalga *Oedogonium intermedium* can meet the nutritional requirements of the herbivorous fish *Ancistrus cirrhosus*," *Algal Res.*, vol. 27, pp. 21–31, 2017.
72. P. W. Glynn et al., "State of corals and coral reefs of the Galápagos Islands (Ecuador): Past, present and future," *Mar. Pollut. Bull.*, vol. 133, pp. 717–733, 2018.
73. L. M. Brander, P. Van Beukering, and H. S. J. Cesar, "The recreational value of coral reefs: A meta-analysis," *Ecol. Econ.*, vol. 63, no. 1, pp. 209–218, 2007.
74. F. Moberg and C. Folke, "Ecological goods and services of coral reef ecosystems," *Ecol. Econ.*, vol. 29, no. 2, pp. 215–233, 1999.
75. G. Tolentino-Pablico, N. Bailly, R. Froese, and C. Elloran, "Seaweeds preferred by herbivorous fishes," in *Nineteenth International Seaweed Symposium*, Springer, Dordrecht, 2007.
76. Y. Zhou, F. Wei, W. Zhang, Z. Guo, and L. Zhang, "Copper bioaccumulation and biokinetic modeling in marine herbivorous fish *Siganus oramin*," *Aquat. Toxicol.*, vol. 196, pp. 61–69, 2018.
77. A. Cheminée et al., "Nursery value of *Cystoseira* forests for Mediterranean rocky reef fishes," *J. Exp. Mar. Bio. Ecol.*, vol. 442, pp. 70–79, 2013.
78. C. S. Ruz, A. F. Muth, F. Tala, and A. Pérez-matus, "The herbivorous fish, *Aplodactylus punctatus*, as a potential facilitator of dispersal of kelp, *Lessonia trabeculata*, in Chile," *J. Exp. Mar. Bio. Ecol.*, vol. 500, pp. 112–119, 2018.
79. W. L. Zemke-white and K. D. Clements, "Chlorophyte and rhodophyte starches as factors in diet choice by marine herbivorous fish," *J. Exp. Mar. Bio. Ecol.*, vol. 240, no. 1, pp. 137–149, 1999.
80. J. Tsuboi, S. Abe, K. Fujimoto, H. Kaeriyama, and D. Ambe, "Exposure of a herbivorous fish to 134Cs and 137Cs from the riverbed following the Fukushima disaster," *J. Environ. Radioact.*, vol. 141, pp. 32–37, 2015.

81. J. M. Park and S. N. Kwak, "Seagrass fish assemblages in the Namhae Island, Korea: The influences of seagrass vegetation and biomass," *J. Sea Res.*, vol. 139, pp. 41–49, 2018.

82. A. L. Bloomfield and B. M. Gillanders, "Fish and invertebrate assemblages in seagrass, mangrove, saltmarsh, and nonvegetated habitats," *Estuaries*, vol. 28, no. 1, pp. 63–77, 2005.

83. T. P. Hughes et al., "Climate change, human impacts, and the resilience of coral reefs," *Science*, vol. 301, no. 5635, pp. 929–934, 2003.

4 The Vertical Farm
Are We There Yet?

Kheir Al-Kodmany

CONTENTS

4.1   INTRODUCTION

4.1.1   BACKGROUND

This study stems from a larger research project that examined vertical-density applications to the Twenty-First Century City [2,3]. As cities try to cope with rapid population growth—adding 2.5 billion dwellers by 2050—and grapple with destructive sprawl, politicians, planners, and architects have become increasingly interested in the vertical-city paradigm. Unfortunately, cities all over the world are grossly unprepared for embracing vertical density because it may aggravate multidimensional sustainability challenges resulting in a "vertical sprawl" that could have worse consequences than "horizontal sprawl." Because of their enormous scale, tall buildings exert significant demand on infrastructure and transportation systems, resulting in unbearable traffic congestions. They also influence the microenvironment by casting shadows and blocking views and sunlight. Tall buildings are expensive to building, operate, and maintain. They also intrude on the existing built environment that matches the human scale. Consequently, they may reduce the overall quality of urban life. One key problem of future cities will be transporting large amount of food to serve dense population, and the vertical farm model offers a potential solution to this problem [1–5].

4.1.2   GOALS AND SCOPE OF THE STUDY

As urban population continues to grow and as arable land is diminishing rapidly across the globe, a fundamental change in food production is needed [4–7]. In particular, building-based urban agriculture is increasingly needed in dense urban environments and a review of current cultivation techniques and projects would likely to contribute positively to academic discussions [6–10]. This is particularly important because vertical farming engages multiple disciplines of natural sciences, architecture, and engineering and affects both people and the environment [9–11]. This chapter attempts to answer the following questions:

- What is a vertical farm?
- What are the driving forces for building it?
- What are the involved high-tech farming methods?
- What are the salient project examples of vertical farming?
- What are the implications for the vertical city?

4.1.3 What Is a Vertical Farm?

In principle, the vertical farm is a simple concept; farm up rather than out [8–10]. The body of literature on the subject distinguishes between three types of vertical farming [9–11]. The first type refers to the construction of tall structures with several levels of growing beds, often lined with artificial lights. This often modestly sized urban farm has been springing up around the world. Many cities have implemented this model in new and old buildings, including warehouses that owners repurposed for agricultural activities [8]. The second type of vertical farming takes place on the rooftops of old and new buildings, atop commercial and residential structures as well as on restaurants and grocery stores [9,10]. The third type of vertical farm is that of the visionary, multistory building. In the past decade, we have seen an increasing number of serious visionary proposals of this type. However, none has been built. It is important, nevertheless, to note the connection between these three types: the success of modestly sized vertical farm projects, and the maturation of their technologies will likely pave the way for the skyscraper farm [9].

4.1.4 Why Vertical Farms?

4.1.4.1 Food Security

Food security has become an increasingly important issue. Demographers anticipate that urban population will dramatically increase in the coming decades. At the same time, land specialists (e.g., agronomists, ecologists, and geographers) warn of rising shortages of farmland [4–6]. For these reasons, food demand could exponentially surpass supply, leading to global famine. The United Nation (UN) estimates that the world's population will increase by 40%, exceeding 9 billion people by the year 2050 [12]. The UN also projects that 80% of the world's population will reside in cities by this time. Further, it predicts that by 2050 we will need 70% more food to meet the demands of 3 billion more inhabitants worldwide [12]. Food prices have already skyrocketed in the past decades, and farmers predict that prices will increase further as oil costs increase and water, energy, and agricultural resources diminish [7–10]. The sprawling fringes of suburban development continue to eat up more and more farmland. On the other hand, urban agriculture has been facing problems as a result of land scarcity and high costs. We desperately need transformative solutions to combat this immense global challenge [8–11].

4.1.4.2   Climate Change

Climate change has contributed to the decrease of arable land. Through flooding, hurricane, storms, and drought, valuable agricultural land has been decreased drastically, thereby damaging the world economy [7,11,12,13,14,18]. For example, as a result of an extended drought in 2011, the United States lost a grain crop assessed at $110 billion [11,19,20]. Scientists predict that climate change and the adverse weather conditions it brings will continue to happen at an increasing rate. These events will lead to the despoliation of large tracts of arable land, rendering them useless for farming. It is common for governments to subsidize traditional farming heavily through mechanisms such as crop insurance from natural causes [6,21,22]. Furthermore, traditional farming requires substantial quantities of fossil fuels to carry out agricultural activities (e.g., plowing, applying fertilizers, seeding, weeding, and harvesting), which amounts to over 20% of all gasoline and diesel fuel consumption in the United States. We need to understand that "food miles" refers to the distance crops travel to reach centralized urban populations. On average, food travels 1500 miles from the farm field to the dinner table [8,15]. In special circumstances—cold weather, for example—food miles can rise drastically as stores, restaurants, and hospitals fly produce in from overseas to meet demands. On a regular basis, more than 90% of the food in major US cities is shipped from outside. A 2008 study at Carnegie Mellon concluded that food delivery is responsible for 0.4 tons of carbon dioxide emissions per household per year [23,24]. This is especially important given the increasing distance between farms and cities from global urbanization. Sadly, the resulting greenhouse gas emissions from food transport and agricultural activities have contributed to climate change (Figure 4.1).

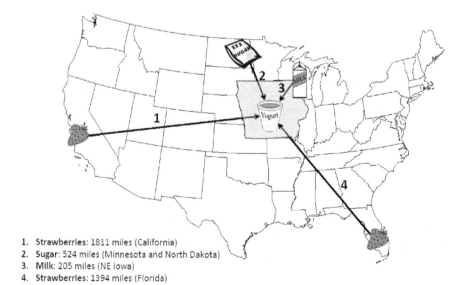

1. **Strawberries:** 1811 miles (California)
2. **Sugar:** 524 miles (Minnesota and North Dakota)
3. **Milk:** 205 miles (NE Iowa)
4. **Strawberries:** 1394 miles (Florida)

FIGURE 4.1   Food travels great distances from farm fields to dinner tables. The map illustrates the case of the travel of basic ingredients of a strawberry yogurt can. (Adapted from Despommier, D., *Trends Biotechnol.*, 31, 388–389, 2013.)

4.1.4.3   Urban Density

Vertical farming offers advantages over "horizontal" urban farming because the former frees land for incorporating more urban activities (i.e., housing more people, services, and amenities) [7]. Research has revealed that designating urban land to farming results in decreased population density, which leads to longer commutes. "If America replaced just 7.9% of its whopping one billion acres of crop and pastureland with urban farms, then metropolitan area densities would be cut in half" [4, p. 71]. Lower density living incurs higher energy use and generates more air and water pollution. The National Highway Travel Survey (NHTS) indicates, "If we decrease urban density by 50%, households will purchase an additional 100 gallons of gas per year. The increased gas consumption resulting from moving a relatively small percentage of farmland into cities would generate an extra 1.77 tons of carbon dioxide per household per year" [23]. Dickson Despommier details space efficiency of vertical farms. He suggested that a 30-story building (about 100 m high) with a basal area of 2.02 ha (5 ac) would be able to produce a crop yield equivalent to 971.2 ha (2400 ac) of conventional horizontal farming. This means that the production of 1 high-rise farm would be equivalent to 480 conventional horizontal farms [24,25].

4.1.4.4   Health

Conventional farming practices often stress profit and commercial gain while paying inadequate attention to inflicted harm on the health of both human and the natural environment [7,8,10]. These practices repeatedly cause erosion, contaminate soil, and generate excessive water waste. Regarding human well-being, the World Health Organization (WHO) has determined that more than half of the world's farms still use raw animal waste as fertilizer, which may attract flies and may contain weed seeds or disease that can be transmitted to plants [1]. Consequently, people's health is adversely affected when they consume such produce. Further, growing crops in a controlled indoor environment would provide the benefit of reducing the excessive use of pesticide and herbicide, which create polluting agricultural runoff [25]. According to Renee Cho, "In a contained environment, pests, pathogens, and weeds have a much harder time infiltrating and destroying crops" [25]. When excess fertilizer washes into water bodies (e.g., rivers, streams, and oceans), a high concentration of nutrients is created (called *eutrophication*), which could disturb the ecological equilibrium. For example, eutrophication may accelerate the proliferation of algae. However, when it dies, microbes consume algae and suck all the oxygen in water, resulting in dead aquatic zones [8]. "As of 2008, there were 405 dead zones around the world" [25]. Further, indoor vertical farming employs high-tech growing methods that use little water (about one-tenth of that used in traditional farming) by offering precision irrigation and efficient scheduling [25,26]. This can have a significant ameliorative effect because demands on water will increase as the urban population grows. Agricultural activities use more than two-thirds of the world's fresh water, and farmers are losing the battle for crop water because urban areas are expanding and consuming more water. The water crisis is likely to become severer as climate change causes warmer temperatures and proliferates more droughts [25].

4.1.4.5 The Ecosystem

Another argument that the vertical farm proponents use is that traditional agriculture has been encroaching on natural ecosystems for millennia. For example, according to Despommier, "Farming has upset more ecological processes than anything else—it is the most destructive process on earth" [4, p. 7]. In the past half-century or so, the Brazilian rainforest has been severely impacted by agricultural encroachment, with some 1,812,992 km^2 (700,000 mi^2) of hardwood forest being cleared for farmland [4]. Despommier suggested that encroachment on these ancient ecosystems is furthering climate change. In this way, indoor vertical farming can reduce the agricultural impact on the world's ecosystems by restoring biodiversity and reducing the negative influences of climate change. If cities employed vertical farms to produce merely 10% of the ground area they consume, this might help to reduce carbon dioxide (CO_2) emissions enough to develop better technological innovations for improving the condition of the biosphere long-term. By eliminating fertilizer runoff, coastal and river water could be restored, and fish stock of wild fish could increase. Wood et al. summarize this point by stating "The best reason to consider converting most food production to vertical farming is the promise of restoring [the] services and functions [of ecosystems]" [26, p. 110].

4.1.4.6 Economics

Proponents of the vertical farm also argue that it will supply competitive food prices [27]. The rising expense of traditional farming is quickly narrowing the cost gap. For example, when vertical farms are located strategically in urban areas, it would be possible to sell produce directly to the consumer, reducing transportation costs by removing the intermediary, which can constitute up to 60% of costs [27]. Vertical farms also use advanced technologies and intensive farming methods that can exponentially increase production. Researchers have been optimizing indoor farming by calibrating, tuning, and adjusting a wide range of variables, including light intensity, light color, space temperature, crop and root, CO_2 contents, soil, water, and air humidity [27–29]. In addition, vertical farming provides an opportunity to support the local economy. Abandoned urban buildings can be converted into vertical farms to provide healthy food in neighborhoods where fresh produce is scarce. Additionally, the high-tech environment of indoor farming can make it fun to farm. Hence, a technology-savvy younger generation has been enticed by the practice, grooming a new breed of farmers. Further, vertical farming provides impetus in the development of innovative agricultural technologies. Finally, it could reconnect city dwellers with nature through the activity of farming [27].

4.2 HIGH-TECH INDOOR FARMING

Progressively, indoor farming relies on most recent advances in technology and sciences. It attempts to take advantage of new machinery and equipment to enable growing greater number of crops in any place at any time. New technologies and innovative farming methods tend also to be efficient in using resources such as water and light, consequently, reducing production costs. They are also increasingly environmentally friendly, abating air, water, and soil pollution. This section reviews and illustrates major methods and technologies involving indoor farming.

4.2.1 Farming Methods

Researchers have advanced myriad methods of urban and vertical farming in the hopes of contributing to sustainable food production. Advanced farming methods could provide greater yields and use far less water than traditional farming [17,18]. The design, layout, and configuration of these high-tech farms would provide optimal light exposure, along with precisely measured nutrients for each plant. Designed to grow in a controlled, closed-loop environment, these farms would eliminate the need for harmful herbicides and pesticides, maximizing nutrition and food value in the process. Indoor farmers could also "engineer" the taste of produce to cater to people's preferences [30]. Researchers intend to develop, refine, and adapt these systems so that they can be ultimately deployed anywhere in the world and provide maximum production and minimum environmental impacts. They represent a paradigm shift in farming and food production and scholars view them as suitable for city farming where land availability is limited [5]. These systems (mainly hydroponics, aeroponics, and aquaponics) and associated technologies are rapidly evolving, diversifying, and improving (Table 4.1). This chapter explains these systems in a gradual manner, from simple to complex.

TABLE 4.1
High-Tech Indoor Farming

Farming Method	Key Characteristics	Major Benefits	Common/Applicable Technologies
Hydroponics	Soilless based, uses water as the growing medium	Fosters rapid plant growth; reduces, even eliminates soil-related cultivation problems; decreases the use of fertilizers or pesticides.	Computerized and monitoring systems; cell phones, laptops, and tablets; food-growing apps; remote control systems and software (farming-from-afar systems); automated racking, stacking systems, moving belts, and tall towers; programmable LED lighting systems; renewable energy applications (solar panels, wind turbines, geothermal, etc.); closed-loop systems, anaerobic digesters; programmable nutrient systems; climate control, HVAC systems; water recirculating and recycling systems; rainwater collectors; insect-killing systems; robots
Aeroponics	A variant of hydroponics; it involves spraying the roots of plants with mist or nutrient solutions.	In addition to benefits mentioned previously, aeroponics requires less water.	
Aquaponics	It integrates aquaculture (fish farming) with hydroponics.	Creates symbiotic relationships between the plants and the fish; it uses the nutrient-rich waste from fish tanks to "fertigate" hydroponics production beds; and hydroponic bed cleans water for fish habitat.	

HVAC = heating, ventilation, and air conditioning.

4.2.1.1 Hydroponics

Hydroponics is a method of growing food using mineral nutrient solutions in water without soil. *Encyclopedia Britannica* defines hydroponics as "the cultivation of plants in nutrient-enriched water, with or without the mechanical support of an inert medium such as sand or gravel" [27, p. 8]. The term is derived from the Greek words *hydro* and *ponos*, which translates to "water doing labor" or "water works." The use of water as a medium for crop growing is not totally new, but the commercial intro-duction of hydroponics arose only recently [28]. National Aeronautics and Space Administration (NASA) researchers have seen hydroponics as a suitable method for growing food in outer space. They have been successful in producing vegetables such as onions, lettuce, and radishes. Overall, researchers have advanced the hydroponic method by making it more productive, reliable, and water-efficient. And, currently, the use of hydroponics in industrial agriculture has become widespread, providing several advantages over traditional soil-based cultivation.

One of the primary advantages of this method is that it could eliminate or at least reduce soil-related cultivation problems (i.e., insects, fungus, and bacteria that grow in soil) [28–30]. The hydroponic method is also relatively low maintenance as well, insofar as weeding, tilling, kneeling, and dirt removal are not issues. The hydroponic method also provides a less labor-intensive way to manage larger areas of produc-tion [8,31,32]. Furthermore, it may offer a cleaner process given that no animal excreta are used. Furthermore, the hydroponic method provides an easier way to con-trol nutrient levels and pH balance. According to Ebba Hedenblad and Marika Olsson, "In soil, many factors, such as temperature, oxygen level, moisture, and microorgan-isms, affect how soil-fixed nutrients are made accessible to plants since the nutrients are being dissolved in water through erosion and mineralization. Therefore, the hydro-ponic method may result in more uniform [produce] and better yields, as the optimum combination of nutrients can be provided to all plants" [30, p. 17].

4.2.1.2 Cylindrical Hydroponic Growing Systems

The Volksgarden or cylindrical Omega Garden hydroponic growing system is a rotating-system technology in which plants are placed inside rotary wheels. When wheels spin, plants rotate around centralized induction lights. The wheels rotate once every 50 minutes using a low-horsepower motor (it is possible to run the wheels via wind turbines and solar panels). In advanced rotary systems, the "plants rotate constantly and slowly around the light source, and their roots pass through a nutrient solution when they reach the bottom of the orbit. Turning at a constant rate allows the plants to take advantage of orbitotropism (based on the impact of gravity on growth) to grow bigger, stronger and faster" [32, pp. 28–29]. The Volksgarden system also provides a compact arrangement for the plants' roots in rock wool, thereby allowing the plants to grow more quickly than in traditional hydroponics [32].

Importantly, the "Ferris wheels" can multiply their capacity by adding "extreme verticality," that is, unit stacking. To appreciate the efficiency of the system, experts have noted, "Each cylinder holds 80 plants, and six cylinders are stacked together about 20 feet high at each station" [32, p. 28]. This adds up to 480 units per station requiring only 3.4 m^2 (36 ft^2) of space. Green Spirit Farms plans to fit 200 stations compactly in one of its vertical farms to grow 96,000 plants per year. For comparison, "conventional basil growers average 16,000 plants per acre (43,560 ft²), less than 20% of the production Green Spirit Farms could have in just 7200 ft²" [32, p. 28]. Furthermore, the Volksgarden system efficiently uses distilled water, requiring one-tenth the water used by traditional hydroponic systems. "Their distillation process allows multiple reuses of water. Rather than discarding the nutrient-dense liquid that remains after the produce has been harvested, it can be re-distilled and reused" [32, p. 29]. Furthermore, the Volksgarden system entails virtually no evaporation because the liquid reservoir for the growing system is closed. Additional water savings are provided by harnessing rainwater, collectively minimizing the demand on municipal water systems [32].

4.2.1.3 Ultrasonic Foggers

Scientists have designed ultrasonic fogger systems to minimize maintenance and maximize yield. They envision using them for myriad horticultural applications, including hydroponics, to provide multiple benefits such as [33]:

- Supplying upper roots with nutrient enriched fogs that penetrate deep into root tissues, keeping them moist, well-nourished, and free of decay [16].
- Promoting the growth of minuscule root hairs, which exponentially increase the root's ability to absorb water, nutrients, and exchange gases [32].
- Reducing the use of water and nutrients by up to 50% [32].
- Reducing the need for bulky and costly growing mediums [33].
- Efficiently using space because the units are compact and designed to be fed by a remotely located reservoir [33].
- By integrating ultrasonic foggers, hydroponic systems come close to aeroponic systems [33].

However, there are some concerns that the hydroponic method relies heavily on chemicals whereby all of the nutrients supplied to the crop are dissolved in water [29]. A hydroponic system is based on chemical formulations to supply concentrations of mineral elements [30]. Liquid hydroponic systems use floating rafts and the Nutrient Film Technique (NFT), and they largely rely on noncirculating water culture—though, new recirculation systems can be applied in NFT techniques [30]. Further, some complain that the produce is tasteless because of all the added chemicals in the system and because the roots do not get adequate oxygen [30]. These shortcomings are partially addressed by the aeroponic method.

4.2.1.4 Aeroponics

Aeroponics is a technological leap forward from traditional hydroponics. An aeroponic system is defined as an enclosed air and water and nutrient ecosystem that fosters rapid plant growth with little water and direct sun and without soil or media [34]. The major difference between hydroponic and aeroponic systems is that the former uses water as the growing medium and the latter has no growing medium. Aeroponics uses mist or nutrient solutions instead of water, so it does not require containers or trays to hold water. It is an effective and efficient way of growing plants because it requires little water (requires 95% less water than traditional farming methods) and needs minimal space [34]. Plant boxes can be stacked up in almost any setting, even a basement or warehouse.

The stacking arrangement of plant boxes is structured so that the top and bottom of the plants are suspended in the air, allowing the crown to grow upward and the roots downward freely. Plants are fed through a fine mist of nutrient-rich, water-mix solution. Because the system is enclosed, the nutrient mix is fully recycled, leading to significant water savings. This method, therefore, is particularly suitable in water-scarce regions. An additional advantage of the aeroponic method is that it is free of fertilizers or pesticides. Furthermore, research has revealed that this high-density planting method makes harvesting easier and provides higher yields. For example, one of the aeroponic experiments with tomato in Brooklyn, New York, resulted in quadrupling the crop over a year instead of the more common one or two crops [34].

4.2.1.5   GrowCube

Recent research and technological development take the aeroponic method to a higher level of productivity and efficiency. For example, GrowCube has proposed a new aeroponic prototype through the high-tech cube, which contains five light plastic plates that spin via a rotisserie-esque wheel and are lit by a strip of light-emitting diodes (LEDs) that provide the necessary light for photosynthesis [34]. At the top of the cube, a device sprays a nutrient-rich mist. The cube and its devices are controlled and managed via computer and software, and sensors inside the cube communicate with the computer to optimize the microclimate. The cube is also pressurized and equipped with an ultraviolet germicidal lamp and a high-efficiency particulate absorption (HEPA) filter, as well as "bug-killing filters in the pipes where the nutrient mixes are pumped" [34].

Consequently, the microclimate inside the cube is bug free, making its produce free of pathogens. Remarkably, IT companies are developing special apps and food-growing food recipes, increasingly available online. Consequently, the aeroponics system and the entire growing process can be optimized remotely [34]. "When it comes time to planting, simply stick your seeds in a growing medium … and download the iOS app. From there, you can select and download a 'grow recipe' from the cloud. … Users are also encouraged to tweak and fork the recipes as they see fit, helping to improve the growing and to offer variations. So if you want crisper lettuce, you can select that as an option" [34]. Furthermore, by conducting the work autonomously, the computer-controlled environment reduces human errors and minimizes the effort of growing food [35].

With such a computerized system, almost anyone could become a sophisticated farmer. What is more, the computerized system will help to "engineer" taste and other characteristics producing crispy or spicy produce! GrowCube has managed to produce "herbs, flowers and foodstuffs like wheatgrass, microgreens, pea-shoots and even 28 heads of lettuce," and it plans to produce fruits such as grapes [34]. The prototype is costly and will likely benefit from economies of scale when it is produced in masses. Consequently, GrowCube plans to expand the project by producing hundreds of these high-tech cubes [34].

4.2.1.6   Solar Aquaculture

Solar aquaculture involves growing high-quality fish protein in small, clean, translucent, and controllable ponds that are exposed to sunlight. Microscopic green algae (nonflowering plants lacking a true stem, roots, and leaves) live in the pond with the fish and grow by absorbing nutrients from the water. In addition, sunlight that strikes the pond helps the algae to grow and causes the water to become warmer. Fish and algae grow faster in warmer water. This method could be suitable for vertical farms, enabling higher rates of production in limited spaces. A solar pond that is 1.5 m high, 1.5 m in diameter (5 ft high, 5 ft diameter) and contains 2649 L (700 gal) of water can produce an annual growth of 18 kg (40 lb) of fish [35].

In addition to supporting fish, solar ponds can serve indirectly as storage units for solar heat. Algae capture about 5% of the entered solar energy, and water absorbs the rest (95%). The pond makes air cooler during the day, given that much of the incoming sunlight is stored as warm water rather than hot air. In contrast, the pond warms the air at night as it releases heat. As such, heat from a solar pond can substitute for heating a greenhouse with gas, oil, or wood or electricity, thereby saving on energy. However, the solar pond requires extensive maintenance because of the fish waste and some of the uneaten food that transforms into waste. These problems are addressed by closed-loop systems and the aquaponics method.

4.2.1.7 Aquaponics

Aquaponics is a biosystem that integrates recirculated aquaculture (fish farming) with hydroponic vegetable, flower, and herb production to create symbiotic relationships between the plants and the fish. It achieves this symbiosis through using the nutrient-rich waste from fish tanks to "fertigate" hydroponic production beds. In turn, the hydroponic beds also function as biofilters that remove gases, acids, and chemicals, such as ammonia, nitrates, and phosphates, from the water. Simultaneously, the gravel beds provide habitats for nitrifying bacteria, which augment the nutrient cycling and filter water. Consequently, the freshly cleansed water can be recirculated into the fish tanks. In one experimental project, aquaponics consisting of wetland pools containing perch and tilapia, whose waste provided nutrients for greens, solved the principal problems of both hydroponics and aquaculture as mentioned previously [36] (Figure 4.2).

Researchers envision that the aquaponic system has the potential to become a model of sustainable food production by achieving the 3Rs (reduce, reuse, and recycle). It offers bountiful benefits, such as [36]:

- Cleaning water for the fish habitat;
- Providing organic liquid fertilizers that enable the healthy growth of plants;
- Providing efficiency because the waste products of one biological system serves as nutrients for a second biological system;
- Saving water because water is reused through biological filtration and recirculation. This feature is attractive particularly in regions that lack water;
- Reducing, even eliminating, the need for chemicals and artificial fertilizers;
- Resulting in a polyculture that increases biodiversity;
- Supplying locally grown healthy food because the only fertility input is fish feed and all of the nutrients go through a biological process;
- Facilitating the creation of local jobs; and
- Creating an appealing business that supplies two unique products—fresh vegetables and fish—from one working unit.

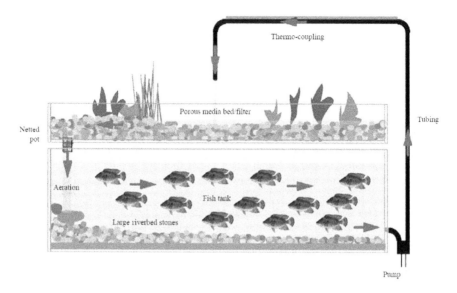

FIGURE 4.2   Basics of an aquaponic system. (Adapted from Martin, G. et al., *Sustainability*, 8, 409, 2016.)

Consequently, aquaponics is preferable to hydroponics. However, aquaponic systems continue to be at the experimental stage, having had limited commercialized success. This is because the technologies necessary to build aquaponic systems are relatively complex, requiring the mutual dependence of two different agricultural products. For this reason, aquaponics also requires intensive management [36].

4.2.2   LIGHTING TECHNOLOGIES

One of the important components of successful vertical farming is sound lighting. Available LED technologies provide only 28% efficiency, an efficiency rate that should be increased to about 50%–60%, at a minimum, to make indoor farming methods cost-effective [37]. Fortunately, experimental developments in LEDs have reached that mark [37,38]. Dutch lighting engineers at Philips have produced LEDs with 68% efficiency. Such an increase in lighting efficiency will dramatically cut costs. Also, a Dutch-based group called PlantLab has recently invented a lighting technology that could help to grow food with a small footprint. According to Michael Levenston, "This invention replaces sunlight with LEDs that produce the optimal wavelength of light for plant growth. Contrary to the sun, traditional assimilation lighting, and TL lighting LED only omits one color of light. No energy is wasted with light spectra that are not used … by the plant" [38]. As such, the new lighting technology provides the correct lighting colors plants need for photosynthesis— blue, red, and infrared light.

Furthermore, new "induction" lighting technology simulates the color spectrum of sunlight to foster the growth of vegetables and fruits. "The light uses an electro-magnet to excite argon gas as its light source, instead of a filament. For this reason, [it] uses much less energy and can last up to 100,000 hours, twice as long as an LED light" [39]. It also generates more heat than LED light but less than an incandescent bulb. Therefore, the lights create enough heat for growing plants without wasting energy to heat the entire building. Moreover, the light units are calibrated to create an "ideal" microenvironment by producing high-quality lighting that is similar to daylight. These units are also long lasting, with a life span of about one decade, and are sold at affordable prices.

4.2.3 FARMING OPERATION

Researchers predict that farming operations will be fully automated in the near future. For example, monitoring systems will be widely implemented (in the form of sensors near each plant bed) to detect a plant's need for water, nutrients, and other requirements for optimal growth and development. Sensors can also warn farmers by signaling the presence of harmful bacteria, viruses, or other microorganisms that cause disease. Also, a gas chromatograph technology will be able to analyze flavonoid levels accurately, providing the optimal time for harvesting. These specific technologies are not totally new. Their development has been ongoing and will likely proliferate in the near future [39].

4.2.4 FARMING FROM AFAR

One of the promising ideas under development is "farming from afar." The cell phone, its software, and apps will ultimately handle much of the day-to-day tending of crops, and vertical farmers will be able to manage multiple farms remotely. New apps will allow farm managers to adjust "nutrient levels and soil pH balance from a smartphone or tablet, and sound alarms if, say, a water pump fails on a vertical-growing system. … So if I'm over in London, where we're looking for a future vertical farm site to serve restaurants, I'll still be able to adjust the process in Michigan or Pennsylvania," as Paul Marks explained [40]. Farming from afar will drastically reduce operational costs by reducing labor and will provide considerable convenience, flexibility, and efficiency in managing farms. Further, by engaging new information technology and working with new online applications, farming could become an exciting and fun activity.

4.2.5 "CLOSED-LOOP AGRICULTURAL" ECOSYSTEMS

"Closed-loop agricultural" ecosystems intend to mimic natural ecosystems that treat waste as a resource. Similar to aquaponics, the waste of one part of the system becomes the nutrients for the other. The closed-loop system recycles and reuses nearly every element of the farming process—dirty water, sewage, and nutrients. Food waste can also be converted to compost. In a closed-loop system, everything remains in the system, leading to a zero-waste outcome. This results not only in drastic decreases in waste but also in the creation of energy and other by-products such as bedding and potting soil.

4.2.5.1   Anaerobic Digester

An anaerobic digester is a biogas recovery system that converts food waste into biogas to produce power and heat [41]. The Plant, a vertical farm in Chicago, has employed an anaerobic digester that captures the methane from 27 tons of daily food waste to produce electricity and heat. Figure 4.3 illustrates how The Plant has integrated an anaerobic digester in its employed close-loop system (also see Section 3.2 on The Plant). Similarly, Great Northern Hydroponics (GNH) in Quebec, Canada, has employed a cogeneration machine that reduces its heating costs and reliance on fossil fuels. GNH's power production has increased such that it is capable of selling electricity back to the Ontario Power Authority, decreasing the province's dependence on fossil fuels.

The main features of the closed-loop systems are as follows:

- At the heart of the system is an anaerobic digester that turns organic materials into biogas, which is piped into turbine generator to make electricity for plant grow light.
- The plants make oxygen to the Kombucha tea brewery, and Kombucha tea brewery makes CO_2 to the plant.
- Waste from the fish feeds the plants and the plants clean the water for the fish.
- More fish waste goes to the digester along with plants' waste, waste from outside sources, and spent grain from the brewery.
- Spent barley from the brewery feeds the fish.
- Sludge from the digester that becomes algae duckweed also feeds the fish.

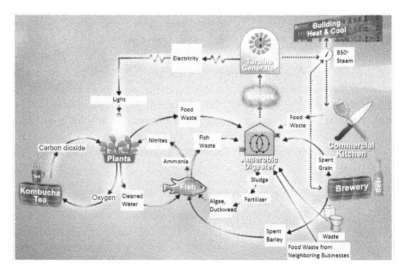

FIGURE 4.3   An illustration of an integrated food production through a closed-loop system. (Adapted from AgSTAR: Biogas Recovery in the Agriculture Sector, *United States Environmental Protection Agency*, Available online: https://www.epa.gov/agstar, accessed on July 15, 2017.)

- Along electricity, the turbine makes steam that is piped to the commercial kitchen, brewery, and the entire building for heating and cooling.
- Therefore, the kitchen produces Kombucha tea, fresh vegetables, fish, beer, and food, all with no waste.

4.2.6 RENEWABLE ENERGY

Some vertical farms have implemented, and others have proposed, employing wind turbines and photovoltaic panels to supply power. Other systems, such as thermal systems that collect solar heat and warehouse refrigeration exhaust, are also under consideration.

4.2.7 INTEGRATION WITHIN CITY INFRASTRUCTURE

Future proposals, for example by Plantagon, envision the integration of vertical farms with the city symbiotically. The proposal envisions that the vertical farm will collect organic waste, manure, CO_2, and excess heat from plants and factories and transform these into biogas for heating and cooling. In this way, the vertical farm not only could grow food but also could help to develop sustainable solutions for better energy, heat, waste, and water use (Figure 4.4).

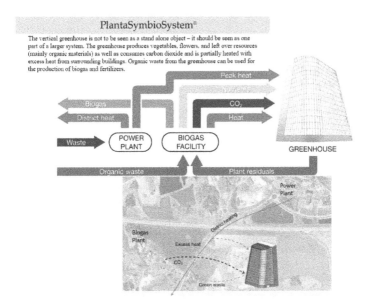

FIGURE 4.4 The proposed vertical farm in the downtown of Linköping, south of the capital Stockholm in Sweden by Plantagon provides an industrial symbiotic system. Partnership will be established among Plantagon, a local energy company, and a local biogas plant. The greenhouse gets district heating from the power plant that runs through a major road. It gets excess heat and carbon dioxide from the biogas plant, and the leftover from the greenhouse goes into the biogas digestor. (Adapted from Advantages of Vertical Farming, *Vertical Farming Systems*, Available online: http://www.verticalfarms.com.au/advantages-vertical-farming, accessed on July 15, 2017.)

4.2.8 REDEFINING VERTICAL FARMS

The aforementioned technologies are redefining the vertical farm as "a revolutionary approach to producing high quantities of nutritious and quality fresh food all year round, without relying on skilled labor, favorable weather, high soil fertility or high water usage" [24]. These new systems add advantages to vertical farming, summarized in Table 4.2.

TABLE 4.2

Advantages of High-Tech Vertical Farming Systems

1. Reliable harvests	Controlled indoor environments are independent of outside weather conditions and would provide consistent and reliable growing cycles to meet delivery schedules and supply contracts.
2. Minimum overheads	Production overheads would decrease by 30%.
Low energy usage	The use of high-efficiency LED lighting technology ensures minimum power use for maximum plant growth. Computer management of photosynthetic wavelengths, in harmony with phase of crop growth, further minimizes energy use while ensuring optimized crop yields.
Low labor costs	Fully automated growing systems with automatic SMS text messaging would require manual labor only for on-site planting, harvesting, and packaging.
Low water usage	Vertical farms would use around 10% of the water required for traditional open field farming.
Reduced washing and processing	Vertical farms would employ strict bio-security procedures to eliminate pests and diseases.
Reduced transport costs	Positioning of facilities close to the point of sale would dramatically decrease travel times, reducing refrigeration, storage, and transport costs in the process.
3. Increased growing areas	Vertical farms would supply nearly ten times more growing area than traditional farms.
4. Maximum crop yield	Irrespective of external conditions, vertical farms can provide more crop rotations per year than open field agriculture and other farming practices. Crop cycles are also faster because of controlled temperature, humidity, light, etc.
5. Wide range of crops	The vertical farm would provide a wide range of crops.
6. Fully integrated technology	The vertical farm would be fully monitored, controlled, and automated.
Optimum air quality	The temperature, carbon dioxide (CO_2), and humidity levels of the vertical farm would be optimized at all times.
Optimum nutrient and mineral quality	The vertical farm would use especially formulated, biologically active nutrients in all crop cycles, providing organic minerals and enzymes to ensure healthy plant growth.
Optimum water quality	All fresh water's contaminants would be removed before entering the vertical farm.
Optimum light quality	High-intensity low-energy LED lighting would be specifically developed and used for maximum growth rates, high reliability, and cost-effective operations.

Source: Advantages of Vertical Farming, *Vertical Farming Systems*, Available online: http://www.verticalfarms.com.au/advantages-vertical-farming, accessed on July 15, 2017.

4.3 VERTICAL FARM PROJECT EXAMPLES

Several cities have embarked on vertical farming projects. The following narrative provides a concise summary, and Table 4.3 offers a list of these projects. The table differentiates among three different types of farming: Low Rise (LR), High Rise (HR), and Rooftop (RT). The LR type refers to 1- to 2-story structure that is used to grow food internally. In contrast, the HR type refers to a tall multistory building that is used to grow food entirely or partially.

4.3.1 MODEST-SCALE VERTICAL FARMS

Companies interested in modest-scale vertical farming are proliferating around the world. For example, Nuvege in Kyoto, Japan, is a 2787 m^2 (30,000 ft^2) hydroponic facility with 5295 m^2 (57,000 ft^2) of vertical grow space that produces a variety of lettuces in a safe environment from the nearby Fukushima nuclear plant [25]. PlantLab in Den Bosch, Holland, is a three-story underground vertical farm that uses advanced LED technology that calibrates light composition and intensity to precise needs, entirely removing the wavelengths of sunlight that prevent plant growth [25]. The farm employs an automated system that monitors and controls numerous variables including humidity, CO_2, light intensity, light color, air velocity, irrigation, nutritional value, and air temperature [25]. The high-tech farm produces a yield three times the amount of the average greenhouse while reducing water use by 90%.

TABLE 4.3
Summary of Examined Projects

Vertical Farm	Location	Type	Status
Nuvege	Kyoto, Japan	LR	Built
PlantLab	Den Bosch, Holland	LR	Built
Sky Greens	Singapore	LR	Built
Green Spirit Farms	New Buffalo, Michigan, USA	LR	Built
FarmedHere	Bedford Park, Illinois, USA	LR	Built
The Plant	Chicago, Illinois, USA	LR	Built
Green Girls Produce	Memphis, Tennessee, USA	LR	Built
Brooklyn Grange	Brooklyn, New York, USA	RT	Built
Gotham Greens	Brooklyn, New York, USA	RT	Built
Plantagon	Sweden	HR	Proposed
La Tour Vivante	France	HR	Proposed
Harvest Green Tower	Vancouver, Canada	HR	Proposed
Skyfarm	Toronto, Canada	HR	Proposed
Pyramid Farm	NA	HR	Proposed
TBD	Philippine	HR	Proposed

HR = High Rise; LR = Low Rise; RT = Rooftop.

4.3.1.1 Sky Greens

One of the promising vertical farms is the Sky Greens of Singapore. As a small island, but with a population of more than five million, Singapore faces potential issues of food security. With land at a premium, limited space for farming is available. Singapore produces only 7% of the food it consumes, and only 250 acres of the island are devoted to farming. The remaining need is supplied by food imports from all over the world. However, the transportation costs of food are becoming increasingly prohibitive. For these reasons, Singapore has been taking vertical farming seriously. Although Singapore is an extreme case, it represents a looming problem facing myriad cities worldwide [26].

Sky Greens is Singapore's first commercial "tropical vegetable urban vertical farm … to achieve enhanced green, sustainable production of safe, fresh and delicious vegetables, using minimal land, water, and energy resources" [43]. The five-year-old farm is three stories tall (9 m or 30 ft) and uses a method called "A-Go-Gro (AGG) Vertical Farming" that uses translucent green houses to grow tropical leafy vegetable year-round at significantly higher yields than traditional farming methods. Sky Greens is capable of producing one ton of fresh veggies every other day. It supplies a variety of tropical vegetables including Chinese cabbage, spinach, lettuce, xiao bai cai, Bayam, kang kong, cai xin, gai lan, and nai bai. By providing high-quality produce at relatively affordable costs, the farm has thrived and intends to expand its production, providing a wider variety of vegetables.

Structurally, the AGG system consists of tall aluminum A-frames that can be as high as 9 m (30 ft) tall with 38 tiers of growing troughs that contain various growing media—soil and hydroponics. The A-frame system takes up only 5.6 m^2 (60 ft^2), making it ten times more efficient than conventional farming [43]. The troughs slowly rotate around the aluminum frame (about three rotations per day) to ensure that the plants obtain uniform sunlight. Such continuous exposure also reduces or even eliminates the need for artificial lighting in some areas of the building. Rotation is powered by a patented low carbon hydraulic system that contains trays of plants. The hydraulic system is an ancient technology empowered with a modern twist; it is a closed-loop that makes efficient use of gravity and consumes little energy. Each 9-m (30 ft) tower uses only 60 W of energy and, therefore, the owner spends only about "$360/month ($3/tower) on electricity" to power the farm [43].

In addition to providing commercial benefits, Sky Greens is engaged in educational programs in the surrounding neighborhoods where students visit the farm, getting exposure and hands-on experience in transplanting, harvesting, and understanding the power of science and technology in creating green urban solutions. According to Sky Green's website, the project provides numerous economic and environmental benefits, summarized in Table 4.4. The project started as a prototype developed jointly with Agri-Food and the Veterinary Authority of Singapore (AVA) in 2010. Sky Greens together with AVA won the Minister for National Development's Research & Development Award 2011 (Merit Award) for vertical farming. Sky Greens promises to become a viable food supplying option [43,44].

TABLE 4.4
The Environmental and Socioeconomic Benefits of the Sky Greens Vertical Farm Project

Environmental Benefits

Environmentally friendly and high tech	Sky Greens observes, learns, and works with nature to achieve sustainability for the good of the environment to grow safe, high-quality vegetables using green technologies.
Low energy usage	Outdoor green houses have abundant sunlight in the tropics. The A-Go-Gro system uses patented low carbon hydraulic green technology to power the rotation of the tower at very-low energy costs, while still allowing the plants to receive abundant sunlight.
Low water usage	As the troughs of plants rotate, irrigation occurs using an innovative flooding method, using very little water. Water is also recycled and reused.
Good waste & water management	Sustainable water management practices are used with all organic wastes being composted at the farm to ensure the use of safe, high-quality fertilizers.
Green technologies	Green technologies have been stringently implemented at the farm to achieve the three Rs (reduce, reuse, and recycle).

Socioeconomic Benefits

Increased productivity	The production yield of Sky Greens Farm is five to ten times greater per unit of area than traditional Singaporean farms that grow leafy vegetables in a conventional fashion.
Tasty vegetables	Tropical leafy vegetables are grown in special soil-based media, which contribute to good-tasting vegetables, suitable for stir-fry and soups. The vegetables are harvested every day and delivered almost immediately to retail outlets for consumers.
Year-round production	As the vertical farm structures are in protected outdoor green houses, the vegetables are grown in a controlled environment, protected from pests, winds, and floods.
Consistent and reliable harvest	A steady supply of fresh leafy vegetables is assured as growing takes place in a controlled environment.
Easy to install and easy to maintain	The modular A-frame rotary system allows quick installation and easy maintenance.
Better ergonomics & automation	The rotary system allows the troughs to be immediately adjusted for easy harvesting. Automation increases the productivity of workers per ton of vegetables grown.
Space savings	The footprint of the vertical system is small but can produce significantly more per unit area than traditional farms. It can also be customized to suit different crop requirements and varying environments.

Source: Sky Greens, Available online: https://www.skygreens.com/, accessed on July 15, 2017.

In the United States, cities such as New York, Chicago, Milwaukee, and others are becoming pioneers of vertical farming by repurposing vacant urban warehouses, derelict buildings, and high-rises to grow food. With so much vacant space available, the cost of property is often affordable to buy or rent. Within the buildings, vertical farmers build tall structures with several levels of growing beds, often lined with artificial lights, to grow crops and "microgreens" (i.e., salad vegetables such as arugula), Swiss chard, mustard, beetroot, and sunflowers. Indoor vertical farming is playing an important role in spurring economic development by repurposing vacant industrial buildings, supplying fresh and healthy food, and providing jobs in distressed areas. Among the pioneering vertical farm projects to spread across the United States are those carried by companies such as Green Spirit Farms, FarmedHere, The Plant, and Green Girls.

4.3.1.2   Green Spirit Farms

Located in New Buffalo, Michigan, Green Spirit Farms (GSF) is a professional food company that has openly embraced vertical farming. The New Buffalo facility has grown out of a former plastic factory. The building contains about 3716 m^2 (40,000 ft^2) of space and sits on an 11-ha (27-ac) site. As standard practice, GSF will enter older vacant industrial or commercial buildings to supply produce nearby urban markets. It aims to provide local markets with high-quality, fresh, pesticide-free, non-genetically modified organism (GMO) foods at affordable prices. The company chooses to grow products with a high local demand like lettuce, basil, spinach, kale, arugula, peppers, tomatoes, stevia, strawberries, and Brussel sprouts. It sells its produce locally to grocery stores and restaurants and to a host of small "Harvest Markets," which sells directly to consumers. GSF runs vertical organic farms in Atlanta, Philadelphia, Canada, and the United Kingdom [45]. The company has a strong belief in vertical farming. According to Green Spirit Farms' Research and Development Manager Daniel Kluko, the future of farming is heading in one clear direction: the vertical. "If we want to feed hungry people this is how we need to farm. … We cut out the risk of traditional farming, the labor, and most of the equipment costs. … This is not a niche business, it's not something novel, this is a necessity for the human race to continue to live" [46].

GSF has advanced several technologies to grow vegetables. These include the Volksgarden Rotary Garden unit, referred to as a Rotary Vertical Growing Station (RVGS), and a multilevel tray system, referred to as a Vertical Growing Station (VGS). GSF has lately commercialized rotary and vertical farming systems using patented techniques to grow local vegetables, herbs, and some fruits and has opened vertical farms in repurposed industrial buildings, including one in East Benton, Pennsylvania. The new facility constitutes a major expansion compared to GSF's first facility in New Buffalo, Michigan, containing 1715 vertical growing stations that will produce herbs, leafy vegetables, peppers, and tomatoes, the equivalent of 81 ha (200 ac) of farmland harvested year-round. This is enabled by facility's efficiency, which uses "98 percent less water, 96 percent less land, and 40 percent less energy" than would be required by traditional agriculture [46]. It is expected that the facility will create more than 100 jobs to support the local economy. GSF has invested about $27 million to establish the vertical farm and received financial aid including a $300,000 Pennsylvania First Program grant, $303,000 in Job Creation Tax Credits, and a $45,450 Guaranteed Free Training grant to train new employees [47]. The location has appealed to Green Spirit because of its proximity to large local markets, with most of its produce selling within approximately 75 miles of the farm [47]. In summary, the vertical farm project provides a useful example of adaptive reuse established through a strong public-private partnership. This has been made possible through the collaboration between GSF and several agencies including the Commonwealth of Pennsylvania, Lackawanna County, Benton Township, and the Greater Scranton Chamber of Commerce.

4.3.1.3 FarmedHere

FarmedHere is a company that was founded in 2011 and has recently expanded to three locations in Illinois: Englewood, a Chicago Southside neighborhood; Flanagan in downstate Illinois; and recently in Bedford Park, a Southwest Chicago suburb. As the company grows, it expects to supply 6% or more of the Chicago area's demand for premium green and culinary herbs. The company also hires local youths through Windy City Harvest, a Chicago Botanic Garden-led urban agriculture-training program targeted to underserved youths. FarmedHere received the U.S Department of Agriculture (USDA) Organic Certification at the end of 2012 [48]. The company's product is spreading in several grocery stores including Whole Foods, Chicago-area Mariano's Fresh Market, Green Grocer, and possibly soon at Trader Joe's and Meijer. FarmedHere was able to receive financial support from Good Food and Whole Foods, the farm's largest customers. The company expects that it has a market niche given the recent generational demands for healthy and organic foods. These new businesses also expect to obtain subsidies from tax-increment financing as well as property-tax breaks for reviving industrial properties [48].

Bedford Park's facility is about 8361 m² (90,000 ft²), much larger than both the first facility in Englewood (371 m² [4000 ft²]) and the second facility in Flanagan (929 m² [10,000 ft²]). Bedford Park's facility, about 24 km (15 mi) from downtown Chicago, is now hyped as the first of its kind and the largest indoor vertical farm in America [48]. It was opened in 2013 and is expected to become a new model for growing produce efficiently in a high-tech manner. The farm resides in a two-story, windowless warehouse and is designed to occupy the full extent of the space. Currently, the farm consists of two structures with large growing beds lit by fluorescent lighting. The first structure contains the aquaponic system in which water circulates between fish tanks, feeding plants that rest in cutouts on Styrofoam "floats" above. The second structure contains the aeroponic system, with water misters underneath that spray the exposed roots of the plants. Workers plant the seeds and grow seedlings on racks that then are transferred into the growing systems. After about a month, the crops are harvested and packaged manually in a cooling room at the facility and then shipped the next morning to grocers in Chicago's metropolitan area [49].

By stacking aquaponic and aeroponic systems vertically, the facility contains 13,935 m² (150,000 ft²) of growing space or about 1.4 ha (3.5 ac). Planting in a controlled environment with ideal humidity and temperature ensures optimal growth. FarmedHere produces about 136,078 kg (300,000 lb) of leafy greens and plans to grow to what will eventually amount to more than 453,592 kg (a million pounds) of chemical-, herbicide-, and pesticide-free leafy greens yearly [49]. It also plans to expand by producing peppers, tomatoes, and other popular vegetables. Their aquaponics produces fish and organic herbs—basil and the like—whereas their aeroponics produces leafy greens like arugula and watercress. For space efficiency, plants are grown on six shelves that receive artificial fluorescent lighting and that are attended by workers using scissor lifts. The aquaponic method filters the nitrogen-rich waste of the tilapia fish and uses it to feed plants, and the hormone-free tilapia are bred in four 3028-L (800-gal) tanks, where water is ultimately recycled to create a closed loop that reduces water use by 97%. Therefore, the system is efficient in its use of water and space.

These new facilities also provide "on-demand farming," meaning they are flexible and responsive to market demands. For example, demand may suddenly increase for particular types of mixed greens or mini greens. "We could change the whole system … and pretty much within the next 14 to 28 days, we [would] have a full grown plant, whatever the market requires" [50]. However, the prime obstacle these farms face remains the electricity needed to grow the plants and heat the space. Because of exorbitant energy bills, some indoor farms have been closed down. In his book *The Vertical Farm: Feeding the World in the 21st Century*, Dickson Despommier stresses the fact that energy remains the primary hurdle [4]. Nevertheless, vertical farmers are trying to find solutions by exploring solar, wind, and methane gas as ways to generate electricity or by supplementing artificial light with natural light through windows and skylights. Other farmers are experimenting with flickering lights sufficient to grow plants with little power.

4.3.1.4   The Plant

Located in the heart of Chicago's derelict stockyards, the almost century old site of The Plant has a long history of food production as a former meatpacking facility and the former home of Peer Foods. The four-story, 8686-m^2 (93,500 ft^2) redbrick warehouse is now set to become a major net-zero vertical farm where the operation is fueled by food waste [51]. The zero-energy facility relies on an on-site Combined Heating and Power (CHP) system that contains a large anaerobic digester that converts food waste into biogas to power, heat, and cool the building. The anaerobic digester captures the methane from 27 tons of food waste daily and 11,000 annually and burns it to produce electricity and heat. The Plant plans to turn the facility into a food business incubator, research lab, and educational and training facility for vertical farming. The building's transformation, which started in 2010, was completed in 2016 [51].

The Plant is currently producing greens, mushrooms, bread, and Kombucha tea. Eventually, the facility will combine a tilapia farm, beer brewery, Kombucha brewery, communal kitchen, an aquaponic system, and green energy production. "We're working to show what truly sustainable food production and economic development looks like by farming inside an old meatpacking facility, incubating small craft food businesses, brewing beer and kombucha, and doing it all using only renewable energy that we make onsite. By connecting outputs of one business to the inputs of another, we are harnessing value from materials that most people would throw away" [51].

The conversion of the space into a vertical farm and food business incubator was partly made possible by a $1.5 million grant from the Illinois Department of Commerce and Economic Opportunity (DCEO) to support the development of a comprehensive renewable energy system [51].

4.3.1.5 Green Girls

Green Girls Produce, a professional food company, is Tennessee's first indoor vertical farm that supplies local restaurants with a year-round fresh produce in an effort to improve the health of Memphians and to fight urban blight. The 60,000-ft^2 facility is located in Memphis' Historic Downtown on the fourth floor of the Emerge Building. Restaurants have a desire for microgreens, which give meals an additional flavoring and pizzazz [52]. "Chefs love them because they make a boring dish pop, they add intense flavor, texture and vivid color. ... On top of that, they are nutritious with up to 40 times the nutrients and vitamins of their mature counterparts" [52]. However, restaurants often refrain from purchasing microgreens given their high costs. Restaurants typically pay about $100 a pound for microgreens. The vertical farm reduces the costs down to below $40 per pound. Green Girls estimates a revenue of about $1 million a year. It supplies affordable microgreens and makes a profit because of efficient technologies provided by automated, recirculating hydroponic systems that require only two employees to run. It is characterized by being clean, efficient in its water use (Green Girls uses 90% less water than conventional farming), and green in its energy use, employing only LED lighting [52].

4.3.2 ROOFTOP FARMING

Rooftop farming simply involves the growing of fruits and vegetable on a rooftop. With a dearth of suitable urban farming land, roofs are increasingly being seen as a plausible space for growing food and a proactive measure in building a sustainable future for cities. Indeed, an abundance of unused rooftop spaces prevails. For example, Honolulu's buildings alone contain more than 1,579,351 m^2 (17,000,000 ft^2) of rooftops. Consequently, in recent years, a great number of rooftop farms have sprung up and some green roofs have been transformed into rooftop farms [53–55].

Overall, converting green roofs into rooftop gardens is a rising trend that aims to "scale up" urban agriculture. Similar to green roofs, rooftop farms are viewed as a necessity for combatting the heat-island effect to mitigate stormwater runoff and to insulate buildings. In addition to these environmental benefits, rooftop farming provides the benefits of supplying the community with fresh produce and promoting modest-scale urban agriculture as well as providing tangible connections to food (see Table 4.5). Among the common vegetables grown on rooftops are kale, collard greens, carrots, radishes, peppers, beans, beets, cherry tomatoes, and various herbs [56]. Though, it should be noted that a slight distinction between a vertical garden and a vertical farm exists. Although both grow plants vertically, the former not always produces fruits and vegetable, and the latter does that exclusively. Vertical farms usually occupy larger areas than that by vertical gardens. Nevertheless, the produce of both types (vertical gardens and vertical farms) is offered to local communities, stores, and restaurants.

TABLE 4.5
Summary of the Benefits of Rooftop Farms

Environmental Benefits

Energy Conservation	The use of green roofs compared to conventional roofing surfaces significantly affects the energy balance within a building. Studies have revealed that green roofs have the potential to reduce a building's energy use by as much as 30%.
Stormwater Management	A green roof absorbs rainwater and helps to prevent sewer system backups and contaminated stormwater overflow. Green roofs can also help to prevent catastrophic environmental events, such as the Ala Wai Canal sewage spill disaster.
Fossil Fuel Reduction	A rooftop farm can grow hyper-local foods. Growing low-food-mile organic produce substantially reduces the fossil fuel consumption associated with the traditional food transportation system.
Global Warming	Green roofs sequester carbon from the atmosphere, lower the levels of carbon dioxide in the air, eliminate the buildup of greenhouse gases, and keep city temperatures cooler by effectively reducing the "Urban Heat Island Effect."
Biodiversity	By replacing inorganic, lifeless roofs with living and thriving green spaces, green roofs support increased biodiversity in urban environments—offering a habitat for a multitude of organisms—from birds to butterflies to countless other beneficial insects.
Environmental Stewardship	Organic rooftop farming protects soil and water from toxic pesticides, herbicides, fungicides, and other dangerous chemicals typically used in conventional farming.

Socioeconomic Benefits

Community	A rooftop farm is a beacon of sustainable community building that creates tangible connections between farmers and consumers. Rooftop farms have the power to do this throughout the city, no matter how scarce or valuable the land.
Local Food Economy	Rooftop farms generate revenue for local farmers and businesses. In this way, rooftop agriculture can also be viewed as an emerging green technology that creates jobs and improves food self-sufficiency by providing organic and hyper-local produce.
Nutrition and health	As an example, the produce of FarmRoof™ is extraordinarily nutritious and healthy. Thanks to a special soil that has been infused with minerals, trace elements, omegas, proteins, and microorganisms, all of their crops are packed with enzymes, antioxidants, nutrients, and minerals.
Aesthetics and Beauty	Supplanting inorganic, lifeless roofs with vibrant greenery, green roofs can beautify cityscapes and balance an otherwise bleak horizon of concrete and tar.

Source: Cerón-Palma, I. et al., *J. Urban Technol.*, 19, 87–103, 2012.

Rooftop farming is not a wholly original idea given that its history goes back to the ziggurats of ancient Mesopotamia and the Hanging Gardens of Babylon. To this day, however, continuing challenges in implementing rooftop farms persist. The structure must be strong enough to support the heavy weight of soil and greenhouse structures. Usually, the edges of rooftops are suitable to support moderate loads, but the central areas may need extra reinforcement [57].

Furthermore, a rooftop requires building access, which imposes logistical issues, liability, weather conditions, and insurance risks. Zoning codes could also be obstacles to obtaining permits. For these reasons, urban farmers are struggling to create efficient farming systems while making a profit. Balancing costs and profits indicates that not every green roof is well suited for farming. For example, in 2012 Local Garden, a rooftop farm that was opened in Vancouver, Canada, was recently closed for economic reasons. Overall, rooftop farming is still a work in progress, but it has great potential as an urban farming system [58].

4.3.2.1 Brooklyn Grange

Several rooftop farms have recently sprung up in New York City, the largest of which is Brooklyn Grange rooftop farm—about one acre in size—and is claimed to be the world's largest rooftop garden [59]. Placed on top of a six-story warehouse that was built in 1919, Brooklyn Garage grows a wide range of "organic produce that includes 40 varieties of tomatoes, peppers, fennel, salad greens, kale, Swiss chard, beans and a variety of delicious root vegetables such as beets, carrots, and radishes, as well as herbs" [42]. Their produce is grown in 19-cm (7.5 in) deep beds with Rooflite soil [59,60]. "Rooflite is a lightweight soil composed of organic matter compost and small porous stones which break down to add trace minerals that are needed for the produce to grow into a healthy and mature state" [60]. Brooklyn Grange is thriving and plans to expand its operation [61].

4.3.2.2 Gotham Greens

Gotham Greens is a 1394 m^2 (15,000 ft^2) facility atop a two-story building in Greenpoint, Brooklyn, New York. Constructed in 2011, it is claimed to be the first rooftop hydroponic commercial farm that uses technologically sophisticated controlled environment agriculture (CEA) in an urban setting in the United States. The facility enjoys unusual farming efficiencies given that it uses less square footage to grow 7–8 times more food than traditional farming, providing produce year-round that is pesticide free [62,63]. Gotham Greens offers the advantage of growing summer vegetables in the winter, enhancing New York City's barren landscape at this time and plans to produce 80–100 tons of pesticide-free, premium-quality lettuce, salad greens, and herbs yearly. Gotham Greens is not only aiming to provide quality produce but also to save on energy by employing advanced computer systems that control heating, cooling, irrigation, and plant nutrition, while using on-site solar photovoltaics. Energy use is reduced further by providing natural ventilation, double-glazing, and thermal insulation that is supplied by the rooftop farm and through high-efficiency pumps and fans. As such, the facility optimizes energy use and consumes less land and water than that of a conventional farm. These energy-saving measures are particularly suitable for New York City given its rapid increase in energy costs. Because of these energy-saving innovations, Gotham Greens will likely be able to reduce its production costs significantly [45,46]. Interestingly, Gotham Greens was the only fresh food supplier in New York during the Sandy Hurricane. This highlights the benefits of protected agriculture in urban areas—particularly as we face climate change—while open-air agriculture can suffer from weather damages [64].

4.3.3 Multistory Farms

As mentioned previously, there have been an increasing number of proposals for multistory farms that remain ideas on the drawing board. This is due, in large part, to the fact that these ideas are not yet economically feasible. However, some companies have taken this endeavor seriously and are on the verge of implementing some of these visionary ideas. Among the pioneering companies is Plantagon.

4.3.3.1 Plantagon

Founded in 2008 in Stockholm and headed by Hans Hassle, Plantagon is a Swedish vertical agriculture company that has flourished through the establishment of offices in cities around the world, including Shanghai, China, and Singapore. Plantagon collaborates with other companies such as SymbioCity and SWECO that are also devoted to finding new methods of vertical farming and in clean technologies (Figure 4.5). It has also established research ties with academic institutions such as Linköping University in Linköping, Sweden, Nanyang Technological University in Singapore, and Tongji University in Shanghai, China. The company participated in Shanghai's World Expo 2010 and won the Globe Sustainability Innovation Award that same year. It also won a Silver Stevie Award for being Europe's "most innovative company" in 2012 [65].

4.3.3.1.1 Organizational Structure

Plantagon has embraced an innovative organizational structure, which they call "companization," that combines two legally bound units, a profit-driven company called Plantagon International AB, and a nonprofit association called Plantagon Non-profit Association [65]. This "companization" aims to unite profit-driven

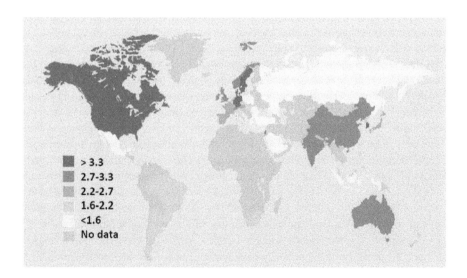

FIGURE 4.5 Map of global cleantech innovation ranking (on a scale from 1 to 5, with 5 being the highest score). (Adapted from Plantagon, Available online: http://plantagon.com/, accessed on July 15, 2017.)

commercial forces with nonprofit organizational values [66]. It is a blend between the top-down action-oriented commercial organization and the bottom-up inclusive and democratic nonprofit association. The two units of the company are dependent on one another, thereby making it difficult for the company to act irresponsibly and unethically. Plantagon is also tied to overarching ethical frameworks provided by the "UN Global Compact" and the "Earth Charter" in their articles of association and statutes. Their board members, from both the profit and nonprofit units, constantly overview financial and social performance, bringing "moral questions" to the otherwise purely economic forum [65,66] (Figure 4.6).

4.3.3.1.2 Technical Innovations

The company is not only innovative in its organizational structure but also in its technology. The company has invented an automated growing food process through a helical structure that would be placed in the center of a building. This structure employs an efficient robotic belt that moves each row of plants one by one rather altogether. At the ground level, workers place seeds in pots and then lifts elevate them to the top of the helix. Here, automatically they are placed on a belt that takes them steadily back down to the ground, controlling the amount of sunlight they receive depending on their age and size. By the time they reach the ground, the plants are ready to be harvested. To keep the crops in sunlight as much as possible, the trays move more quickly when in the shade. Eventually, LED lights will be employed to complement solar exposure (Figure 4.7). Plantagon has also made advancements in hydroponics by introducing pumice soil as a growing medium instead of relying completely on water. Pumice, a volcanic rock that results from the cooling of lava in water, possesses a unique porosity that absorbs nutrients and then channels them to plants. This may overcome the common problem of tastelessness that hydroponically generated produce is notorious for, though both methods provide similar levels of nutrition [66].

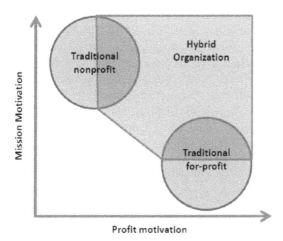

FIGURE 4.6 The concept of "companization" by Plantagon. (Adapted from Plantagon, Available online: http://plantagon.com/, accessed on July 15, 2017.)

FIGURE 4.7 A proposed vertical farm by Plantagon. It comprises a helix structure placed in the center of a sphere-shape building and stretches vertically. The helix structure contains a robot belt that instead of moving thousands of plants over the whole belt, it moves each raw of plants one by one. Seeds are placed in pots on the belt at the ground-floor level. Then plants are elevated to the top by an elevator and placed on the moving belt, which takes them back to the ground, and by the time they reach the ground floor, plants are ready to harvest. (Drawing by author.)

Plantagon offers three approaches to vertical farming: façade farming, multifunctional farming, and standalone farming [65].

4.3.3.1.2.1 The PlantaWall Façade System This approach proposes turning a building's façade into a productive greenhouse. It specifically suggests placing a 6-m-deep (20 ft) greenhouse at the perimeter where solar exposure is greatest. Plants sit in trays that constantly move on parallel conveyor belts, obtaining maximum exposure to natural light in the process. The PlantaWall façade system is based on a flexible modular structure that can be attached to the whole facade or part of an existing structure. In addition to structural flexibility, the façade system provides the benefit of soundproofing, supporting a healthy work environment, and improving thermal insulation and shade. Furthermore, the system fosters a symbiotic relationship that creates a healthy environment for everyone by transferring CO_2 from people to plants and oxygen from plants to people. However, the façade system reduces the amount of natural light that can penetrate deeply into a space [65].

4.3.3.1.2.2 Multifunctional Vertical Farm In addition to farming, the multifunctional vertical farm incorporates functions such as office space, hotel, and retail space as well as residential and educational uses.

4.3.3.1.2.3 Stand-Alone Vertical Farm The stand-alone vertical farm is dedicated exclusively to the industrial production of food. The company has produced two prototypes, a sphere for tropical climates and a half-moon shape for temperate climates.

The latter prototype, named "Plantscraper" is a building that has been proposed in downtown Linköping, south of the Swedish capital of Stockholm (Figure 4.8). The purpose of this structure has been to create a reference building that can be used as a model of vertical farming. The "Plantscraper" is a twelve-story, mixed-used tower that houses an indoor farm along the southern façade (The PlantaWall), a farmers market at the ground floor and office spaces for proposed urban farming research [65]. The renting of office space will provide supplemental income to the building. It is estimated that the Plantscraper will produce between 300 to 500 metric tons of leafy greens, particularly pak choi, a year. Pak choi, also known as celery cabbage, is a Chinese vegetable that can be eaten raw or cooked [65]. This leafy green has been emphasized because Plantagon wants to use this building as a model for Asian cities. In addition, the building will collect and reuse all wastewater, with

FIGURE 4.8 One of Plantagon's prototype for the vertical farm. Seeds are planted in the first floor, then carried to the top via elevators, and next moves down by the helix belt system. (Adapted from Plantagon, Available online: http://plantagon.com/, accessed on July 15, 2017.)

all pesticides, fertilizers, and soil pollution being automatically monitored and controlled. Interestingly, Plantagon plans to integrate the vertical farm with the city's civic infrastructure (i.e., electricity, gas, water, and sewage) [66].

4.3.3.2   La Tour Vivante

The French architecture firm Atelier SOA has proposed a thirty-floor mixed-use vertical farm—Tour Vivante (Living Tower)—that houses farming, residences, office spaces, restaurants, and retail. The concept evokes the sense of a self-contained vertical neighborhood that enjoys autonomy and reduces travel needs between the rural and the urban by bringing together the activities of production and consumption into one space [67]. People who live and work in the tower can enjoy proximity to fresh, ripe, and preservative-free produce, and the tower's tenants provide a constant income that supports the farms. It is also possible that residents of the tower could be employed in the building to reduce travel times between home and work. Additionally, combining farming, housing, and office spaces brings together the traditionally distant rural farming experience close to residents, reconnecting them with nature. The tower's architects explain that "the separation between city and countryside, urban planning and natural areas, places of living, consumption, and production is increasingly problematic for sustainable land management. ... The concept of Tour Vivante aims to combine agricultural production, housing and activities in a single system" [67].

The interweaving of these functions and activities aims to provide a symbiotic relationship between the inhabitants and the farming environment. For example, food waste from restaurants and residents will be collected and processed to be used as a liquid fertilizer to fruits and vegetables. Similarly, the oxygen produced by plants will be channeled to tenants, whereas the CO_2 produced by the tenants will be transferred to plants. Rainwater from the roof and the façades will be collected, filtered, and used in the farm, and the waste generated by the farm and other functions (housing, offices) will be collected and used to generate power for the tower [67].

Blackwater produced by the tower will be recycled and purified to feed and fertilize the plants. The Tour Vivante also embraces renewable energy by employing two large wind turbines, as well as photovoltaic panels, placed on the southern façade of the building and its roof. The tower will also be constructed from recycled and recyclable materials. A thermal concrete shaft at the core of the building will help to control solar gain and humidity in the building while facilitating natural ventilation through the "chimney effect" and the application of a double skin façade. Overall, the tower explores possibilities of blending architecture with agriculture, merging food production and consumption into one place [67].

4.3.3.3   Harvest Green Tower

The city of Vancouver, British Columbia, Canada, expects to face rapid urban growth in the coming decades that would result in substantial demands on food. Romses Architects has recently proposed a multifunctional vertical farm in Vancouver in support of the city's 2030 challenge of reducing carbon emissions. The project, named "Harvest Green Tower," provides space for a mix of uses including residences, offices, entertainment, retail, restaurants, and vertical farming that will produce vegetables, herbs, fruits, fish, egg-laying chickens, goat cheese, and sheep dairy. The project also connects well with the city's public mass-transit system. The main sections of the tower are described as follows [68].

- *Underground.* A parking lot and shared car co-op.
- *Street level.* A grocery store, farmer's market, restaurants, and a transit hub.
- *Lower floors.* A livestock grazing plain, bird habitat, goat cheese, and sheep dairy facility.
- *Middle floors.* A space for producing fruits, vegetables, and fish.
- *Upper floors.* Residential units.
- *Tower's top.* A large rainwater cistern.

The tower will harness renewable energy by installing rooftop mounted wind turbines on the roof, photovoltaic glazing on the façade, a geothermal station underground, and bioenergy from the compost of plants and animals. All of these combined factors provide the possibility of selling surplus power back to the grid. Rainwater will be collected in a cistern atop the tower and used to irrigate plants through the downward pull of gravity. The project was a winning entry in Vancouver's 2030 challenge given that it addressed climate change through the promotion of high-density mixed-use developments in the urban core, thereby curtailing sprawl [68].

4.3.3.4 Skyfarm

Designed by architect Gordon Graff, the Skyfarm is a fifty-nine-floor vertical farm proposed for downtown Toronto, Canada. This tower embraces the hydroponic method for growing food, using an area totaling about 743,224 m^2 (8,000,000 ft^2). It is predicted that the Skyfarm will produce the equivalent of a thousand-acre rural farm, feeding about 35,000–50,000 people per year [69]. The building will be equipped with its biogas plant that contains an anaerobic digester to produce methane from its waste, which is then burned to generate electricity. The system can also reclaim the waste and sewage by diverting it to the Skyfarm's anaerobic digester to produce the methane necessary to generate electricity. Furthermore, the liquid slurry extracted by the anaerobic digesters could be used by rural farms as fertilizer [69].

4.3.3.5 Pyramid Farm

Architecture professor Eric Ellingsen and Microbiology and Public Health professor Dickson Despommier have proposed a self-sustaining thirty-story "Pyramid Farm" to produce a wide variety of fruits and vegetables as well as fish and poultry farm that will feed about 50,000 people annually. The project embraces a closed-loop agricultural ecosystem where nearly every element of the farming process—including water and nutrients—is recycled and reused, minimizing waste drastically. The farm will include a heating and pressurization system that splits sewage into water and carbon, fueling machinery and electrifying lighting. Overall, the proposed farm will be efficient because it will use only 10% of the water used in traditional farming techniques and only 5% of the space of a traditional farm. Finally, the visual quality of the pyramid is appealing and would fit well into an urban area [70].

4.3.3.6 Vertical Farm in Philippine

A recent visionary project proposed a vertical aeroponic farm that offers high yields of rice in the Philippines, a country that faces food insecurity and lacks farmland. Rice is a particularly valuable food not only in the Philippines but also across the globe, given that half of the world's population relies on rice as a major food source. Jin Ho Kim has proposed using aeroponics technology that uses minimal water to grow rice compactly on a terraced vertical farm to be constructed by an array of bamboo parallelograms. The project will create local jobs, supply local food, and lead to significant savings in transportation, storage, refrigeration, and packaging, eventually incorporating spaces for social gatherings and children [70,71].

4.4 DISCUSSION: OPPORTUNITIES AND CHALLENGES

Vertical farming represents a proactive thinking approach that aims to ensure the sustainability of cities by addressing the issue of food security. The urban population already faces food shortages, and food prices are skyrocketing due to increases in oil prices, shortages of water, and the diminishment of other agricultural resources. The current practices of supplying food to urban areas suffer from environmental and economic problems, such as the inefficient practice of transporting food great distances. As an answer to these problems, the vertical farm will grow food efficiently and sustainably by saving energy, water, and fossil fuels, reducing toxins and restoring ecosystems, as well as providing new opportunities for employment. We have seen the rapid growth of modest-scale vertical farming, and these projects have provided excellent examples of adaptive reuse of vacant industrial spaces [72–76].

Therefore, the vertical farm may offer opportunities in the three pillars of sustainability: environment, society, and economy [Table 4.6]. It can offer a sustainable food-production model that supplies crop year-round with no interruption because of climate change, season, or adverse natural events (e.g., hurricane, drought, and flood). It has also the potential to provide greater yield per space unit—the ratio is 1:4–6, depending on the type of crop [77,78]. Further, the high-tech cultivation methods of the vertical farm reduce demand on potable water. They are often efficient in irrigating plants by targeting plant roots and reducing evaporation [79]. They may also recycle wastewater (grey, even black water) and harness rainwater. When fish farms are integrated, fish removes waste (especially fish filet). The vertical farm can also produce energy by burning methane from compost. For example, the Plant Vertical Farm in Chicago and the Republic of South Korea VF factory convert waste to energy [80,81].

When compared to traditional farming, the vertical farm may reduce the need for fossil fuel required for tractors, plows, or shipping. Traditional farming uses lots of fossil fuel; for example, conventional farming in North America consumes 20% of fossil fuel due to plowing, seeding, harvesting, fertilizing and so on [82,83]. The vertical farm can also reduce food travel distance (food miles) by promoting "local for local" life style (i.e., distances between food production and consumption are minimized) [84–86]. As mentioned previously, in conventional farming, food travels on average 1500 miles. Further, the vertical farm eliminates the need of packing agricultural crops for long-distance transportation [6,87,88].

TABLE 4.6
Key Sustainable Benefits of the Vertical Farm

#	Benefit	Environmental	Social	Economic
1	Reducing food-miles (travel distances)	Reducing air pollution	Improving air quality improves environmental and people's health. Customers receive "fresher" local food	Reduce energy, packaging, and fuel to transport food
2	Reducing water consumption for food production by using high-tech irrigation systems and recycling methods	Reducing surface water run off of traditional farms	Making potable water available to more people	Reduce costs
3	Recycling organic waste	Save the environment by reducing needed landfills	Improve food quality and subsequently consumers' health	Turn waste into asset
4	Creating local jobs	People do not have to commute to work and hence will decrease ecological footprint	Create a local community of workers and social networks with farmers	Benefit local people economically
5	Reduced fertilizers, herbicides, and pesticides	Improve the environmental well-being	Improve food quality and subsequently consumers' health	Decrease costs
6	Improve productivity	Needs less space	Reduce redundant, repetitive work, and save time to do productive and socially rewarding activities	Offer greater yields
7	Avoid crop losses due to floods, droughts, hurricane, overexposure to sun, and seasonal changes	Decrease environmental damage and cleanups of farms after damage	Improve food security	Avoiding economic loss
8	Control product and produce regardless to seasons	Produce regarding season	Increase accessibility year-round and improve respond to population demand	Fuel economic activities year-round

(Continued)

TABLE 4.6 (Continued)
Key Sustainable Benefits of the Vertical Farm

#	Benefit	Environmental	Social	Economic
9	Using renewable energy	Reducing fossil fuel	Improve air quality	Reduce costs
10	Bringing nature closer to city	Increase biodiversity	Improve health, reduce stress, and enhance psychological well-being	Create jobs in the city
11	Promoting high-tech and green industry	"Green technology" reduces harm and improves environmental performance	Encourages higher education and generates skilled workers	Provides new jobs in engineering, biochemistry, biotechnology, construction and maintenance, and research and development
12	Reducing the activities of traditional farming	Preserving natural ecological system	Improve health of citizens	Saving money required to correct environmental damage
13	Repurposing dilapidated buildings	Enhances the environment. Removes eyesores and stigma from neighborhoods	Creates opportunities for social interaction	Revive economy

Indoor farming is immune to weather change, which affects traditional farming by changing temperature, water supply, and photo intensity. These factors often reduce produce yield; for example, droughts destroy crops every year worldwide [88,89]. As such, the vertical farm will be important for food security especially as climate change threatens our cities. As mentioned previously, Gotham Greens was the only fresh food supplier in New York during the Sandy Hurricane. Additional benefit of the vertical farm is providing an ideal growth environment for each plant that improves crop yield [62,73]. Advances in technologies, for example, the LED lighting, promises to increase yields as LED emits programmed wavelength of light for optimal photosynthesis of different types of crops. Luckily, the prices of these technologies are dropping [74]. The vertical farm provides an environment almost free of invasive pest species [90]. It also reduces, and possibly eliminates, the use of mineral fertilizers, herbicides and pesticides, and nitrogen (N) and phosphorous (P), which have been causing environmental degradation by polluting surface water and groundwater [91–93].

The vertical farm can assist in cooling the environment, sequestering CO_2, reducing the urban heat island (UHI) effect, and combating climate change [94,95]. As such, it can help in reducing energy needed to cool indoor spaces in summer time and reduce the emission of CO_2 [96–98]. Further, the vertical farm can help in absorbing noise because vegetation reduces sound reflection. Vegetation and soil can function together as sound insulator [97,99]. When rooftop farming or green roofs are also applied, noise could be lowered further by absorbing higher frequency noise produced by auto traffic, machineries, and airplanes [99].

The vertical farm may also provide socioeconomic benefits by offering employment opportunities [100]. Building a vertical farm requires a multidisciplinary team of architects, engineers, scientists, farmers, horticulturists, environmentalists, marketers, and economists. For example, industrial, mechanical and electrical engineers will be needed to design water recycling systems, lighting systems, heating, ventilation, and air conditioning (HVAC) systems, and seed- and plant-growth monitoring and harvesting systems. Computer experts will be needed to build databases and software applications. As such, the vertical farm offers new exciting careers in biochemistry, biotechnology, construction, maintenance, marketing, engineering, and research and development opportunities for improving the involved technologies [100–102]. Further, robotics and software engineers could also be needed. Moreover, a vertical farm may include grocery stores and engage distribution centers, which provide additional work opportunities [68,103].

Further, vertical farming is likely to create new social networks and communities that forge new friendship in the workspace and beyond, among producers, farmers, and consumers [93]. In addition, vertical farmers may enjoy selling their produce directly to customers and develop friendships [104]. Vertical farms could have an important educational role in informing about plants and produce by bringing farming activities closer to city dwellers. For example, Gotham Greens in New York frequently invites visitors and students to their vertical farm and holds educational sessions [105].

Our health is directly impacted by the "freshness" and wholesomeness of food we consume and the vertical farm intends to supply quality, local organic food [58,89]. It could help consequently to reduce or stop the transmission of harmful infectious diseases because currently much produced food by conventional agriculture is polluted and carry bacterial diseases that endanger the lives of millions of inhabitants.

That is, since the vertical farm product is not soil-based, it is likely to be not affected by polluted soil or irrigation water. Further, vertical farm's crops are rich in nutrients [106,107]. Moreover, being close to nature helps to reduce stress and has positive influence on mental health.

Notwithstanding the promising future and large potential benefits, challenges, and barriers in the path to the vertical farm implementation should be noted. Research has highlighted social resistance, where masses of people do not accept the alteration of traditional farming because it is the natural way to grow food [77,89,94,99]. Importantly, the core argument against vertical farming is that growing food indoors requires more energy, effort, and resources than traditional farming [86,98,100]. That is, "It is much more expensive, of course, to build a vertical greenhouse than to build a normal greenhouse" [72]. Despommier acknowledges that the costs of implementing vertical farms are high, particularly the start-up costs, and he calls on the government to provide the seed money to fund these projects. Apparently, to raise the required investment capital is a challenge. In short, for the vertical farm to be sustainable, it must be profitable.

Further, central cities are ideal locations for vertical farms for enjoying proximity to dense population and major retail outlets [69]. However, the issue of affordability is salient in central cities where land and space are expensive. For example, central areas in major cities in Hong Kong, Australia (e.g., Melbourne, Sydney), United Kingdom (e.g., London) have expensive real estate, which presents an economic difficulty at the commercial scale [107]. However, some major cities such as New York and Chicago have sizable stocks of vacant older properties that could be repurposed into vertical farming. This has happened in projects such as The Plant in Chicago, Illinois, and AeroFarms in Newark, New Jersey [49,72].

Overall, residential, retail, office, and commercial uses of high-rises continue to be more profitable than that by agricultural activities [95]. It seems that increasing the productivity of the vertical farm is the prime factor to make it prevail in the future. "If the yield per hectare for indoor farming is much higher than rural outdoor farming, perhaps as much as up to 50 times, this factor will eventually outweigh the initial cost of land acquisition … and assuming 50-fold improved productivity, the break-even point may well be an estimated 6–7 years" [106, p. 295]. Such production will likely to offset the startup costs including expensive land or rent. Another drawback of the vertical farm is inability to produce all types of crops. In fact, current vertical farms produce limited crops such as lettuce, tomatoes, strawberries and, to less extent, grape, and soy products. Also, produced quantities are too small. Martin's research indicates an imbalance between vertical farms' production and their catchment areas, where many of population reside in these areas are underserve. They found that in the near future, urban food will continue to come from distant rural areas [98,107].

Further, because of economic reasons, most vertical farms produce and then distribute leafy greens to restaurants, and local residents remain not the prime client. In the same manner, low-value agricultural commodities such as wheat continue to be economically unviable. Therefore, the current product of vertical farms is limited in scope and quantities. Overall, production volumes of vertical farms are small, particularly when compared to "limitless" acres of traditional farming. Further, scaling up vertical farming could be costly and complex [86,109,110].

Another limitation is that current renewable energy sources, such as photovoltaics and wind turbine, produce little energy that would make it difficult not to rely on the city grid. It is only the plants at the building's perimeter and on the top level that could benefit from solar radiation [34,109]. In this regard, it is important to employ rotatable stacked arrays of plants inside each floor of a high-rise enclosure so that plants receive maximum natural light [93].

Consequently, until now, no multistory vertical farm tower has been built, and despite copious attempts to make this a reality, the concept of the vertical farm tower remains on the drawing board. Despommier hints at a solution by stating that "High-rise food-producing buildings will succeed only if they function by mimicking ecological process, namely by safely and efficiently re-cycling everything organic, and recycling water from human waste disposal plants, turning it back into drinking water" [4, p. 121].

4.5 CONCLUSIONS

A plethora of research and pioneering projects has demonstrated the potential of vertical farming at the pilot scale, prototypes, and at the production levels. The vertical farm has the potential to play a critical role in the sustainability of food in urban areas. This is most important as we project into the future when urban population will increase significantly. Vertical farming has various advantages over rural farming, observed within the three pillars of sustainability: environmental, social, and economic. New high-tech cultivation methods, including hydroponics, aeropnics, and aquaponics, largely challenge the need for soil-based farming for a range of crops. Advancements in greenhouse and supporting technologies such as multi-racking mechanized systems, recycling systems, LED lighting, solar power, wind power, storage batteries, drones as well as computing power, software applications, databases and the Internet of Things are likely to coalesce into efficient production systems in the near future. Increasingly, there is a need for interdisciplinary research and collaboration that promote collective thinking among the various disciplines involved in creating vertical farms [111–114].

Perhaps, in the distant future, there is the prospect of developing fully automated vertical farms. And hypothetically, if vertical farms were integrated in the city, they will be able to supply food for the entire population. However, there is still a need for more developments that scale up projects so that the economic and commercial feasibility and return on investment (ROI) are offered at best rates. As such, there is a need for research that accurately assesses the ROI of various types and sizes of vertical farms. There is a need to investigate the full life-cycle analysis (LCA) and the number of years to reach parity with a traditional farm [109,115,116].

The success of the vertical farm will depend not only on innovation in technologies but also on local conditions including demand on certain produce by population, availability of labors, and farming conditions. An effective organizational structure and sound leadership are also important factors. Creativity, stewardship, and inventiveness are critical ingredients for companies that venture into new businesses such as vertical farming. In a globalized world, competition is stiff, but the first to succeed may gain a competitive edge [99,108,115]. As such, robust and resilient business models are needed in a world characterized by increasing complexity, nonlinearity, and "glocal" exchanges of goods. According to Copenhagen Institute for Futures Studies, Instituttet for Fremtidsforskning "Tomorrow's innovative leader isn't necessarily the person in front with innovative ideas, but the one who discovers the front-runners and harvests their ideas to cultivate and nourish the innovative environment in his organization" [74]. Nevertheless, interest in vertical farming will increase as climate change prevails further and available arable land per capita declines [117].

One more serious obstacle remains; the increasing populations of developing countries. Do these countries possess the required technologies and technical expertise to implement the vertical farm? These countries are largely poor. Can we make the products of the vertical farm affordable to the poor? Furthermore, many of these poor populations live in slums, in food deserts, away from modern life. How can we make the produce of vertical farms accessible to slum populations? Ultimately, the effectiveness of vertical farming will depend on various local factors, including the demand and supply of food, urban populations and their density, technological development, culture and eating habits, water and energy supply, as well as weather conditions.

4.6 FUTURE RESEARCH

Vertical farming is growing rapidly, and this research barely scratches the surface of a long and complex endeavor. The examined projects offer catalysts to further developments. Future studies may examine other projects including, but are not limited to, Green Sense Farms (Portage, Indiana, and Shenzhen, China), AeroFarms (Newark, New Jersey), Metropolis Farms (Philadelphia, Pennsylvania), Plenty (San Francisco, California) VerticalHarvest (Jackson, Wyoming), Lufa Farms (Montreal, Quebec, Canada), VertiCrop™ (Vancouver, British Columbia, Canada), and a new unnamed project in Suwon, South Korea. Also, future research may examine specialized technologies and methods for various indoor farming systems. For example, hydroponic systems offer multiple methods, including NFT, Wick System, Water Culture, Ebb and Flow (Flood and Drain), Drip Feed System, and Aeroponic Systems. Further, there is a need for conducting quantitative research that gives accurate assessments of the benefits and shortcomings of various types of vertical farms. Importantly, future research should examine the issue of affordability of advanced equipment of vertical farming to developing countries. Researchers should invent, advance, and further develop local farming techniques to make vertical farm projects feasible in these countries. For example, they may invent recycling methods that reduce reliance on water, design local systems by capturing rainwater, and may capitalize on local solar power for providing natural light and energy.

REFERENCES

1. Al-Kodmany, K. The vertical farm: A review of developments and implications for the vertical city. *Buildings*, 2018, *8*, 24.
2. Al-Kodmany, K. *Eco-Towers: Sustainable Cities in the Sky*; WIT Press: Southampton, UK, 2015.
3. Al-Kodmany, K.; Ali, M.M. *The Future of the City: Tall Buildings and Urban Design*; WIT Press: Southampton, UK, 2013.

4. Corvalan, C.; Hales, S.; McMichael, A.J. *Ecosystems and Human Well-Being: Health Synthesis*; World Health Organization: Geneva, 2005.
5. Healy, R.G.; Rosenberg, J.S. *Land Use and the States*; Routledge: New York, 2013.
6. Thomaier, S.; Specht, K.; Henckel, D.; Dierich, A.; Siebert, R.; Freisinger, U.B.; Sawicka, M. Farming in and on urban buildings: Present practice and specific novelties of Zero-Acreage Farming (ZFarming). *Renew. Agric. Food Syst.* 2015, *30*, 43–54.
7. Despommier, D. *The Vertical Farm: Feeding the World in the 21st Century*; Thomas Dunne Books: New York, 2010.
8. Despommier, D. Farming up the city: The rise of urban vertical farms. *Trends Biotechnol.* 2013, *31*, 388–389.
9. Despommier, D. *Encyclopedia of Food and Agricultural Ethics (Vertical Farms in Horticulture)*; Springer: Dordrecht, the Netherlands, 2014.
10. Touliatos, D.; Dodd, I.C.; McAinsh, M. Vertical farming increases lettuce yield per unit area compared to conventional horizontal hydroponics. *Food Energy Secur.* 2016, *5*, 184–191.
11. Muller, A.; Ferré, M.; Engel, S.; Gattinger, A.; Holzkämper, A.; Huber, R.; Müller, M.; Six, J. Can soil-less crop production be a sustainable option for soil conservation and future agriculture? *Land Use Policy* 2017, *69*, 102–105.
12. The United Nations. *World Population Prospects: The 2017 Revision*; United Nations: New York, 2017.
13. Mukherji, N.; Morales, A. *Zoning for Urban Agriculture. Zoning Practice 3*; American Planning Association: Chicago, 2010.
14. Katz, R.; Bradley, J. *The Metropolitan Revolution. How Cities and Metropolitan Areas Are Fixing Broken Politics and Fragile Economy*; The Brookings Institution: Washington, DC, 2013.
15. Astee, L.Y.; Kishnani, N.T. Building integrated agriculture: Utilising rooftops for sustainable food crop cultivation in Singapore. *J. Green Build.* 2010, *5*, 105–113.
16. Säumel, I.; Kotsyuk, I.; Hölscher, M.; Lenkereit, C.; Weber, F.; Kowarik, I. How healthy is urban horticulture in high traffic areas? Trace metal concentrations in vegetable crops from plantings within inner city neighbourhoods in Berlin, Germany. *Environ. Pollut.* 2012, *165*, 124–132.
17. Kalantari, F.; Tahir, O.M.; Joni, R.A.; Fatemi, E. Opportunities and challenges in sustainability of vertical farming: A review. *J. Landsc. Ecol.* 2017, *2*, 2.
18. Kalantari, F.; Tahir, O.M.; Lahijani, A.; Kalantari, S. A review of vertical farming technology: A guide for implementation of building integrated agriculture in cities. *Adv. Eng. Forum* 2017, *24*, 76–91.
19. Martin, G.; Clift, R.; Christie, I. Urban cultivation and its contributions to sustainability: Nibbles of food but oodles of social capital. *Sustainability* 2016, *8*, 409, doi:10.3390/su8050409.
20. United States Department of Agriculture. Food Desert Locator. Available online: https://www.fns.usda.gov/tags/food-desert-locator (accessed on July 15, 2017).
21. Padmavathy, A.; Poyyamoli, G. Enumeration of arthropods in context to plant diversity and agricultural (organic and conventional) management systems. *Int. J. Agric. Res.* 2016, *6*, 805–818.
22. Sanyé-Mengual, E.; Cerón-Palma, I.; Oliver-Solà, J.; Montero, J.I.; Rieradevall, J. Environmental analysis of the logistics of agricultural products from roof top greenhouses in Mediterranean urban areas. *J. Sci. Food Agric.* 2013, *93*, 100–109.
23. Blaustein-Rejto, D. Harvard economist claims urban farms do more harm than good. *Inhabitat*, June 24, 2011. Available online: http://inhabitat.com/harvard-economist-claims-urban-farms-do-more-harm-than-good/ (accessed on July 15, 2017).
24. Food and Agriculture Organization (FAO). *Good Agricultural Practices for Greenhouse Vegetable Crops: Principles for Mediterranean Climate Areas*; FAO: Roma, Italy, 2013; Chapter 15.

25. Cho, R. Vertical farms: From vision to reality. *State of the Planet, Blogs from the Earth Institute*, October 13, 2011. Available online: http://blogs.ei.columbia.edu/2011/10/13/vertical-farms-from-vision-to-reality/comment-page-1/ (accessed on June 1, 2014).

26. Wood, S.; Sebastian, K.; Scherr, S.J. *Pilot Analysis of Global Ecosystems: Agroecosystems*; International Food Policy Research Institute and World Resources Institute: Washington, DC, p. 110, 2001. Available online: http://www.wri.org/publication/pilot-analysis-global-ecosystems-agroecosystems (accessed on July 15, 2017).

27. Al-Kodmany, K. Sustainable tall buildings: Cases from the global south. *Int. J. Archit. Res.*, 2016, *10*(2), 52–66.

28. Harris, D. *Hydroponics: A Practical Guide for the Soilless Grower*, 2nd ed.; New Holland Publishing: London, 1992.

29. Munoz, H.; Joseph, J. Hydropnics: Home-based vegetable production system, Inter-American Institute for Cooperation on Agriculture (IICA). June, 2010. Available online: https://repositorio.iica.int/bitstream/handle/11324/3156/BVE17089162i.pdf;jsessionid=182B2B2E863C1E9539C35E55AE87E695?sequence=1 (accessed on July 15, 2017).

30. Hedenblad, E.; Olsson, M. Urban growth analysis of crop consumption and development of a conceptual design to increase consumer adoption of vertical greenhouses. Master Thesis, Chalmers University of Technology, Gothenburg, Sweden, 2017. Available online: http://www.tekniskdesign.se/download/Hedenblad_Olsson.pdf (accessed on July 15, 2017).

31. Pullano, G. Indoor vertical grower touts concept's benefits. *VGN, Vegetable Grower News*, August 15, 2013. Available online: http://vegetablegrowersnews.com/index.php/magazine/article/indoor-vertical-grower-touts-concepts-benefits (accessed on July 15, 2017).

32. Green Spirit Farms. Sustainable vertical farming. Available online: http://www.green-spiritfarms.com/in-the-news (accessed July 15, 2017).

33. Yeang, K. Ecoskyscrapers and ecomimesis: New tall building typologies. In *Proceedings of the 8th CTBUH World Congress on Tall & Green: Typology for a Sustainable Urban Future*, Healthcare City, Dubai, March 3–5, 2008; pp. 84–94.

34. Cooper, D. GrowCube promises to grow food with ease indoors (hands-on), *Engaget*, November 8, 2013. Available online: http://www.engadget.com/2013/11/08/insert-coin-growcubes-hands-on/ (accessed on July 15, 2017).

35. Barnhart, E. A primer on New Alchemy's Solar Aquaculture, December, 2017. Available online: http://www.aces.edu/dept/fisheries/education/documents/Primeron solaraquaculture_aquaponics.pdf (accessed on July 15, 2017).

36. Diver, S. Aquaponics—Integration of hydroponics with aquaculture, National Sustainable Agriculture Information Service, 2006. Available online: http://www.back-yardaquaponics.com/Travis/aquaponic.pdf (accessed on July 15, 2017).

37. Eve, L. PlantLab could grow fruit and vegetables for the entire world in a space smaller than Holland, *Inhabitat*, March 17, 2015. Available online: http://inhabitat.com/dutch-company-plantlabs-agricultural-revolution-could-grow-the-worlds-fruit-and-veg-in-a-space-smaller-than-holland/ (accessed July 15, 2017).

38. Levenston, M. Philips lighting promotes city farming, *City Farmer News*, December 10, 2011. Available online: http://www.cityfarmer.info/2011/12/10/ (accessed on July 15, 2017).

39. Matuszak, J. Vertical farming revolution taking root in New Buffalo, *Harbor Country News*, July 3, 2012. Available online: http://www.harborcountry-news.com/articles/2012/07/04/features/doc4ff35c9fc8e3d166588244.txt (accessed on July 15, 2017).

40. Marks, P. Vertical farms sprouting all over the world, *New Scientist*, January 16, 2014. Available online: http://www.newscientist.com/article/mg22129524.100-vertical-farms-sprouting-all-over-the-world.html#.U1yPU_RDuao (accessed on July 15, 2017).

41. AgSTAR: Biogas Recovery in the Agriculture Sector, *United States Environmental Protection Agency*. Available online: https://www.epa.gov/agstar (accessed on July 15, 2017).

42. Advantages of Vertical Farming, *Vertical Farming Systems*, 2017. Available online: http://www.verticalfarms.com.au/advantages-vertical-farming (accessed on July 15, 2017).
43. Sky Greens. Available online: https://www.skygreens.com/ (accessed on July 15, 2017).
44. Aiken, M. Vertical farming powering urban food sources, *Diplomatic Courier*, April 3, 2014. Available online: http://www.diplomaticourier.com/news/topics/environment/2143-vertical-farming-powering-urban-food-sources (accessed on July 15, 2017).
45. Green Spirit Farms. Sustainable vertical farming, New Buffalo, MI. Available online: http://www.greenspiritfarms.com/ (accessed on July 15, 2017).
46. Smiechowski, J. Vertical farming venture achieves sustainability and success in New Buffalo, MI, *SeedStock*, June 10, 2013. Available online: http://seedstock.com/2013/06/10/vertical-farming-venture-achieves-sustainability-and-success-in-new-buffalo-michigan/ (accessed on July 15, 2017).
47. Frank, L. Pennsylvania Governor Corbett Partners with innovative farm to establish operations in Lackawanna County, Creating 101 Jobs, December 13, 2013. Available online: http://www.prnewswire.com/news-releases/pennsylvania-governor-corbett-partners-with-innovative-farm-to-establish-operations-in-lackawanna-county-creating-101-jobs-235605661.html (accessed on July 15, 2017).
48. Trotter, G. FarmedHere, indoor farm in Bedford Park, turning off the lights for good, Chicago Tribune, January 16, 2017. Available online: http://www.chicagotribune.com/business/ct-farmedhere-closing-0117-biz-20170116-story.html (accessed on July 15, 2017).
49. Meinhold, B. FarmedHere: The nation's largest indoor organic farm now growing in Chicago, *Inhabitat*, May 27, 2013. Available online: http://inhabitat.com/farmedhere-the-nations-largest-indoor-organic-farm-now-growing-in-chicago/ (accessed on July 15, 2017).
50. In a Chicago Suburb, an Indoor Farm Goes 'Mega', Associate Press (PA), March 28, 2013. Available online: http://www.cleveland.com/business/index.ssf/2013/03/in_a_chicago_suburb_an_indoor.html (accessed on July 15, 2017).
51. The Plant. Available online: http://www.plantchicago.com/ (accessed on July 15, 2017).
52. Baker, S. Green girl to create indoor vertical farm, *Daily News East Memphis*, November 7, 2012. Available online: http://www.memphisdailynews.com/news/2012/nov/7/green-girl-to-create-indoor-vertical-farm/ (accessed on July 15, 2017).
53. Cerón-Palma, I.; Sanyé-Mengual, E.; Oliver-Solà, J.; Montero, J.I.; Rieradevall, J. Barriers and opportunities regarding the implementation of rooftop eco. Greenhouses (RTEG) in Mediterranean Cities of Europe. *J. Urban Technol.* 2012, *19*, 87–103, doi:10.1080/10630732.2012.717685.
54. Sanyé-Mengual, E.; Antón, A.; Oliver-Solà, J.; Montero, J.I.; Rieradevall, J. Environmental assessment of urban horticulture structures: Implementing rooftop greenhouses in Mediterranean cities. In *Proceedings of the LCA Food Conference*, San Francisco, CA, October 8–10, 2014.
55. Whittinghill, L.J.; Rowe, D.B.; Cregg, B.M. Evaluation of vegetable production on extensive green roofs. *Agroecol. Sustain. Food Syst.* 2013, *37*, 465–484, doi:10.1080/21683565.2012.756847.
56. Orsini, F.; Gasperi, D.; Marchetti, L.; Piovene, C.; Draghetti, S.; Ramazzotti, S.; Bazzocchi, G.; Gianquinto, G. Exploring the production capacity of rooftop gardens (RTGs) in urban agriculture: The potential impact on food and nutrition security, biodiversity and other ecosystem services in the city of Bologna. *Food Secur.* 2014, *6*, 781–792, doi:10.1007/s12571-014-0389-6.
57. Al-Kodmany, K. Green retrofitting skyscrapers: A review. *Buildings*, 2014, *4*, 683–710.
58. Sanyé-Mengual, E.; Llorach-Massana, P.; Sanjuan-Delmás, D.; Oliver-Solà, J.; Josa, A.; Montero, J.I.; Rieradevall, J. The ICTA-ICP Rooftop Greenhouse Lab (RTG-Lab): Closing metabolic flows (energy, water, CO_2) through integrated Rooftop Greenhouses. In *6th AESOP Sustainable Food Planning Conference*; Roggema, R., Keefer, G., Eds.; VHL University of Applied Sciences: Utrecht, the Netherlands, 2014; pp. 692–701.

59. Plakias, A.C. *The Farm on the Roof: What Brooklyn Grange Taught Us about Entrepreneurship, Community, and Growing a Sustainable Business*; Avery: New York, 2016.
60. Brooklyn Grange. Available online: http://brooklyngrangefarm.com/ (accessed on July 15, 2017).
61. Leahy, K. Brooklyn Grange is the world's largest rooftop farm! *Inhabitat*, October 18, 2011. Available online: http://inhabitat.com/nyc/brooklyn-grange-worlds-largest-rooftop-farm-kicks-off-second-growing-season/ (accessed on July 15, 2017).
62. Gotham Greens. Available online: http://gothamgreens.com/ (accessed on July 15, 2017).
63. Something Unexpected Is Sprouting in Historic Chicago Neighborhood, CBS News. Available online: https://www.cbsnews.com/news/gotham-greens-viraj-puri-greenhouse-farm-on-roof-of-pullman-factory-chicago/ (accessed on April 29, 2016).
64. Robin Plaskoff Horton. New York City urban farms after Hurricane Sandy, urban gardens. Available online: http://www.urbangardensweb.com/2012/11/12/new-york-city-urban-farms-after-hurricane-sandy/ (accessed on November 12, 2012).
65. Plantagon. Available online: http://plantagon.com/ (accessed on July 15, 2017).
66. Geddes, T. The future of vertical farming in 5 inspiring examples, dispatch weekly, October 12, 2016. Available online: http://dispatchweekly.com/2016/10/future-vertical-farming-5-inspiring-examples/ (accessed on July 15, 2017).
67. La Tour Vivante, an International Sustainable City, SOA Architects. Available online: http://www.ateliersoa.fr/verticalfarm_fr/pages/images/press_urban_farm.pdf (accessed on July 15, 2017).
68. Jordana, S. Harvest Green Project/Romses Architects, *ArchDaily*, May 7, 2009. Available online: http://www.archdaily.com/21555/harvest-green-project-romses-architects/ (accessed on July 15, 2017).
69. Alter, L. Sky farm proposed for Downtown Toronto, TreeHugger, June 14, 2007. Available online: http://www.treehugger.com/sustainable-product-design/sky-farm-proposed-for-downtown-toronto.html (accessed on July 15, 2017).
70. Kain, A. Pyramid farm is a vision of vertical agriculture for 2060, *Inhabitat*, June 3, 2009. Available online: http://inhabitat.com/pyramid-farm-vertical-agriculture-for-2060/ (accessed on July 15, 2017).
71. Meinhold, B. Aeroponic vertical farm: High-yield terraced rice paddies for the Philippines, *Inhabitat*, March 18, 2013. Available online: http://inhabitat.com/aeroponic-vertical-farm-high-yield-terraced-rice-paddies-for-the-philipines/ (accessed on July 15, 2017).
72. Fletcher, O. The future of agriculture may be up, *The Wall Street Journal*, October 13, 2012. Available online: http://online.wsj.com/news/articles/SB10000872396390443855804577602960672985508 (accessed on July 15, 2017).
73. Despommier, D. The vertical essay. Available online: http://www.verticalfarm.com/?page_id=36 (accessed on July 15, 2017).
74. How to be Resilient in the 21st Century—The Radar, the Shield and the Sword, Copenhagen Institute for Futures Studies, Instituttet for Fremtidsforskning, 2012. Available online: http://cifs.dk/publications/members-reports/ (accessed on July 15, 2017).
75. Albajes, R.; Cantero-Martínez, C.; Capell, T.; Christou, P.; Farre, A.; Galceran, J.; López-Gatius, F. et al. Building bridges: An integrated strategy for sustainable food production throughout the value chain. *Mol. Breed.* 2013, *32*, 743–770.
76. Sivamani, S.; Bae, N.; Cho, Y. A smart service model based on ubiquitous sensor networks using vertical farm ontology. *Int. J. Distrib. Sens. Netw.* 2013, *9*, 461–495.
77. Abel, C. The vertical garden city: Towards a new urban topology. *CTBUH J.* 2010, *2*, 20–30.
78. Eigenbrod, C.; Gruda, N. Urban vegetable for food security in cities: A review. *Agron. Sustain. Dev.* 2015, *35*, 483–498.

79. Safikhani, T.; Abdullah, A.M.; Ossen, D.R.; Baharvand, M. A review of energy characteristic of vertical greenery systems. *Renew. Sustain. Energy Rev.* 2014, *40*, 450–462.
80. Sivamani, S.; Bae, N.-J.; Shin, C.-S.; Park, J.-W.; Cho, Y.-Y. An OWL-based ontology model for intelligent service in vertical farm. *Lect. Notes Electr. Eng.* 2014, *279*, 327–332.
81. Lehmann, S. *The Principles of Green Urbanism: Transforming the City for Sustainability*; Earthscan: London, 2010.
82. Caplow, T. Building integrated agriculture: Philosophy and practice. *Urban Futur.* 2009, *2030*, 54–58.
83. Despommier, D. The vertical farm: Controlled environment agriculture carried out in tall buildings would create greater food safety and security for large urban populations. *J. für Verbraucherschutz und Leb.* 2011, *6*, 233–236.
84. Cicekli, M.; Barlas, N.T. Transformation of today greenhouses into high technology vertical farming systems for metropolitan regions. *J. Environ. Prot. Ecol.* 2014, *15*, 1779–1785.
85. Grewal, S.S.; Grewal, P.S. Can cities become self-reliant in food? *Cities* 2012, *29*, 1–11.
86. Voss, P.M. Vertical farming: An agricultural revolution on the rise. Master's Thesis, Halmstad University, Halmstad, Sweden, 2013.
87. Besthorn, F.H. Vertical farming: Social work and sustainable urban agriculture in an age of global food crises. *Aust. Soc. Work* 2013, *66*, 187–203.
88. Sauerborn, J. Skyfarming: An alternative to horizontal croplands. *Resour. Eng. Technol. A Sustain. World* 2011, *18*, 19.
89. Wagner, C.G. Vertical farming: An idea whose time has come back. *Futurist* 2010, *44*, 68–69.
90. Graber, A.; Schoenborn, A.; Junge, R. Closing water, nutrient and energy cycles within cities by urban farms for fish and vegetable production. *Int. Water Assoc. Newsl.* 2011, *37*, 37–41.
91. Dubbeling, M. Integrating urban agriculture in the urban landscape. *Urban Agric. Mag.* 2011, *25*, 43–46.
92. Sivamani, S.; Kwak, K.; Cho, Y. A rule based event-driven control service for vertical farm system. In *In Future Information Technology*; Park, J.J., Stojmenovic, I., Choi, M., Xhafa, F., Eds.; Springer: Berlin/Heidelberg, Germany, 2014; Volume 276, pp. 915–920.
93. Specht, K.; Siebert, R.; Thomaier, S.; Freisinger, U.; Sawicka, M.; Dierich, A.; Henckel, D.; Busse, M. Zero-Acreage farming in the city of Berlin: An aggregated stakeholder perspective on potential benefits and challenges. *Sustainability* 2015, *7*, 4511–4523.
94. Specht, K.; Siebert, R.; Hartmann, I.; Freisinger, U.B.; Sawicka, M.; Werner, A.; Thomaier, S.; Henckel, D.; Walk, H.; Dierich, A. Urban agriculture of the future: An overview of sustainability aspects of food production in and on buildings. *Agric. Hum. Values* 2014, *31*, 33–51, doi:10.1007/s10460-013-9448-4.
95. World Agriculture: Towards 2015/2030. An FAO perspective, FAO Corporate Document Repository, August 21, 2017. Available online: http://www.fao.org/docrep/005/y4252e/y4252e06.htm (accessed on July 15, 2017).
96. Banerjee, C.; Adenaeuer, L. Up, up and away! The economics of vertical farming. *J. Agric. Stud.* 2014, *2*, 40.
97. Al-Chalabi, M. Vertical farming: Skyscraper sustainability? *Sustain. Cities Soc.* 2015, *18*, 74–77.
98. Ellis, J. Agricultural transparency: Reconnecting urban centres with food production. Master's Thesis, Dalhousie University, Halifax, NS, 2012.
99. Kadir, M.Z.A.A.; Rafeeu, Y. A review on factors for maximizing solar fraction under wet climate environment in Malaysia. *Renew. Sustain. Energy Rev.* 2010, *14*, 2243–2248.

100. Saadatian, O.; Lim, C.H.; Sopian, K.; Salleh, E. A state of the art review of solar walls: Concepts and applications. *J. Build. Phys.* 2013, *37*, 55–79.

101. Glaser, J.A. Green chemistry with nanocatalysts. *Clean Technol. Environ. Policy* 2012, *14*, 513–520.

102. Perez, V.M. Study of the sustainability issue of food production using vertical farm methods in an urban environment within the state of Indiana. Master's Thesis, Purdue University, West Lafayette, IN, 2014.

103. Al-Kodmany, K. Sustainable tall buildings: Toward a comprehensive design approach. *International Journal of Sustainable Design*, 2012, *2*(1), 1–23.

104. Germer, J.; Sauerborn, J.; Asch, F.; de Boer, J.; Schreiber, J.; Weber, G.; Müller, J. Skyfarming an ecological innovation to enhance global food security. *J. für Verbraucherschutz und Leb.* 2011, *6*, 237–251.

105. Al-Kodmany, K. Guidelines for tall buildings development. *Int. J. High-Rise Build.* 2012, *1*(4), 255–269.

106. La Rosa, D.; Barbarossa, L.; Privitera, R.; Martinico, F. Agriculture and the city: A method for sustainable planning of new forms of agriculture in urban contexts. *Land Use Policy* 2014, *41*, 290–303.

107. Ali, M.M.; Al-Kodmany, K. Tall buildings and urban habitat of the 21st century: A global perspective. *Build. J.* 2012, *2*(4), 384–423.

108. Benke, K.; Tomkins, B. Future food-production systems: Vertical farming and controlled-environment agriculture. *Sustain. Sci. Pract. Policy* 2017, *13*, 13–26.

109. Kim, H.-G.; Park, D.-H.; Chowdhury, O.R.; Shin, C.-S.; Cho, Y.-Y.; Park, J.-W. Location-based intelligent robot management service model using RGPSi with AoA for vertical farm. *Lect. Notes Electr. Eng.* 2014, *279*, 309–314.

110. Lam, S.O. Urban agriculture in Kingston: Present and future potential for re-localization and sustainability. Master's Thesis, Queen's University, Kingston, ON, Canada, 2007.

111. Liu, X. Design of a modified shipping container as modular unit for the Minimally Structured & Modular Vertical Farm (MSM-VF). Master's Thesis, The University of Arizona, Tucson, AZ, 2014.

112. Al-Kodmany, K. The logic of vertical density: Tall buildings in the 21st century city. *Int. J. High-Rise Build.* 2012, *1*(2), 131–148.

113. Al-Kodmany, K. Tall buildings, design, and technology: Visions for the twenty-first century city. *J. Urban Technol.* 2011, *18*(3), 113–138.

114. Al-Kodmany, K. Eco-iconic skyscrapers: Review of new design approaches. *Int. J. Sustain. Des.* 2010, *1*(3), 314–334.

115. Tan, Z.; Lau, K.K.-L.; Ng, E. Urban tree design approaches for mitigating daytime urban heat island effects in a high-density urban environment. *Energy Build.* 2015, doi:10.1016/j.enbuild.2015.06.031.

116. Nochian, A.; Mohd Tahir, O.; Maulan, S.; Rakhshandehroo, M. A comprehensive public open space categorization using classification system for sustainable development of public open spaces. ALAM CIPTA, *Int. J. Sustain. Trop. Des. Res. Pract.* 2015, *8*(spec. 1), 29–40.

117. Kalantari, F.; Mohd Tahir, O.; Golkar, N.; Ismail, N.A. Socio-cultural development of Tajan Riverfront, Sari, Iran. *Adv. Environ. Biol.* 2015, *9*, 386–392.

5 Geometry Optimization of a Piezoelectric Microcantilever Energy Harvester

Khalil Khanafer, Ali Al-Masri, and Kambiz Vafai

CONTENTS

NOMENCLATURE

C Piezoelectric coefficient
$d31$ Piezoelectric material coefficient
e Piezoelectric coefficient in strain form
E Elasticity
P Polarization
PZT Piezoelectric
X Design parameter
t_p Thickness of piezoelectric material
V Volume

GREEK SYMBOLS

δ Deflection
ν Poisson Ratio
$\sigma_{\max}$ maximum Stress

5.1   INTRODUCTION

The harvesting of energy from vibrating structures in applications related to powering low-energy electric devices has established a lot of attention in the recent years [1–15]. A power-harvesting device aims at capturing the dissipated energy that is surrounding a system and converting it into useful energy for the consumption of electrical devices. One source of normally dissipated energy is the ambient vibrations existing around most machines and biological systems. This source of energy is best for the use of piezoelectric materials, which have the ability to convert mechanical energy into electrical energy and vice versa. Piezoelectric materials are physically deformed in the presence of an electric field and, on the contrary, produce an electrical charge when deformed under mechanical load. Piezoelectric energy harvesters are suitable for a variety of wireless sensor technologies, such as structural health monitoring [16], agriculture [17], health care [18], and civil and military applications [19–21].

A wireless sensor network (WSN), which consists of many devices communicating wirelessly to each other through several nodes, is designed to achieve sensing, data acquisition, local processing, and wireless communication. These sensors are normally comprised of many parts such as radio transceiver, power sources, and microcontroller [22]. Figure 5.1 illustrates the main subsystems of the wireless sensor node [23]. The typical distribution of power consumption among these subsystems is depicted in Figure 5.2 [23]. It is clearly shown in Figure 5.2 that the communication system, which is composed mainly of a short range radio transceiver with the amplifiers, consumed the largest power compared with other systems. Therefore, it is essential to turn the radio off completely when it is not receiving or transmitting signals.

Conventionally, batteries are used as the power sources for wireless sensors and other embedded electronics. However, these batteries have a limited life and are not reliable as a long-term source of energy, and this exhibits a major effect on the performance of WSNs [24,25]. Therefore, the idea of using piezoelectric material for harvesting energy in overcoming the current challenges presented by the short-term life span of batteries is promising. Energy harvesting by piezoelectric materials was

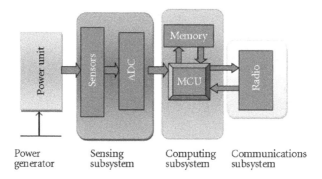

FIGURE 5.1   Architecture of the wireless sensor node. (From Akyildiz, W., et al., *Comput. Netw.*, 38, 393–422, 2002. With permission.)

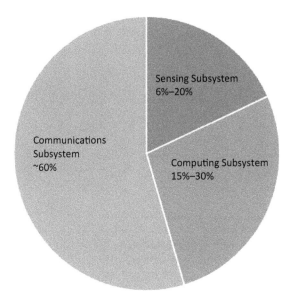

FIGURE 5.2 Wireless sensor nodes power consumption distribution. (From Akyildiz, W., et al., *Comput. Netw.*, 38, 393–422, 2002. With permission.)

studied by many researchers in the past few decades [26–29]. Hausler and Stein [27] led the early studies that investigated the possibility of generating energy from the expansion and contraction of the rib cage during breathing. Their constructed proto-type, which was implemented in vivo on a mongrel dog, was able to produce a peak voltage of 18 V and a power output of 17 µW. Another early study was carried out by Umeda et al. [28] who investigated the piezoelectric (PZT) applications and sug-gested a mechanism for converting the impact energy into electric energy by using a piezoelectric device.

Energy harvesting from human motion has received a lot of attention by research-ers in the literature. Kymissis et al. [30] proposed parasitic energy harvesting while walking from piezoelectric shoes using a unimorph strip made from piezoceramic composite material. It was reported in a previous study that 67 W are available from heel strikes during a brisk walk for an average weight of a person of 68 kg, 2 steps/s, and a heel moving of 5 cm [30]. Qian et al. [32] proposed an embedded piezoelectric footwear harvester for energy generation from human motion (Figure 5.3). Two heel-shaped aluminum plates were used to collect and transfer the heel-strike force to the amplification frames, which in turn transmitted to piezoelectric stacks. The results of that study indicated that the maximum power outputs increased with an increase in the walking speed (Figure 5.4). Further, it was found that more power maybe gen-erated with less number of piezoelectric stacks because of the large force input/stack. For a walking speed of 3.5 mph, the power out for eight stacks harvester was 9 mW/show, and it was 14 mW/show for six stacks harvester.

Hwang et al. [32] developed a self-powered cardiac pacemaker empowered by a flexible piezoelectric energy harvester. The lead magnesium niobate-lead titanate

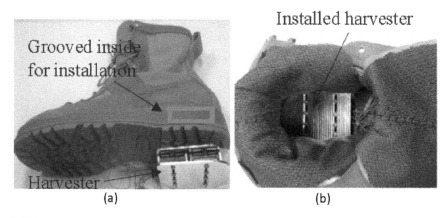

FIGURE 5.3   (a) Assembled harvester. (b) Installed harvester in the heel. (From Starner, T., *IBM Syst. J.*, 35, 618–629, 1996. With permission.)

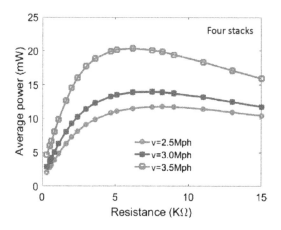

FIGURE 5.4   Average power of the piezoelectric harvester for various walking speeds. (From Starner, T., *IBM Syst. J.*, 35, 618–629, 1996. With permission.)

(PMN-PT) piezoelectric energy harvester was able to generate more than enough power to stimulate the heart without an external power source (Figure 5.5). Yang et al. [34] conducted a preliminary study to convert the mechanical energy of cars into electrical energy using pavement piezoelectric energy harvester. A stacked array type of piezoelectric energy harvester (PEH) was designed with protection package to enhance the characteristics and life span of the PEH. Their preliminary results indicated that the generated energy using PEH was able to light LED signs. Furthermore, as the speed of the car increases, the electric energy generated by PEH increases as depicted in Figure 5.6. Figure 5.6 illustrates that square cross section of PEH exhibits higher peak voltage compared with circular cross section of PEH.

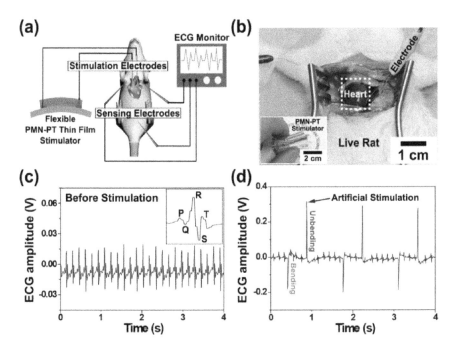

FIGURE 5.5 (a) Artificial cardiac pacemaker using lead magnesium niobate-lead titanate (PMN-PT) energy harvester, (b) Medical experiment on a living rat for heart stimulation, (c) Recorded electrocardiogram (ECG) in a normal rat heart before stimulation, and (d) Recorded ECG after artificial stimulation using PMN-PT flexible piezoelectric energy harvester. (From Qian, F., et al., *Energy Convers. Manage.*, 171, 1352–1364, 2018. With permission.)

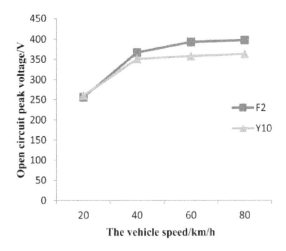

FIGURE 5.6 The open circuit peak voltage of F2 (square cross section of PEH) and Y10 (circular cross section of PEH) at different speeds. (From Hwang, G.T. et al., *Adv. Mater.*, 23, 4880–4887, 2014. With permission.)

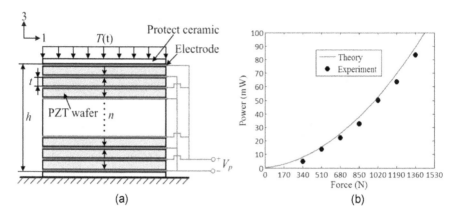

FIGURE 5.7 (a) Schematic of the piezoelectric stack. (b) Maximum output power compared with exciting force amplitude. (From Yang, H., et al., *Int. J. Pavement Res. Technol.*, 11, 168–175, 2018. With permission.)

Jiang et al. [35] developed a compression-based energy harvester that can be embedded into pavement to generate electrical energy from traffic-induced vibration. An experimental testing was conducted to analyze the output power of a harvesting unit consisting of three piezoelectric multilayer stacks (Figure 5.7a). Very good correlation was obtained between experimental and theoretical results (Figure 5.7b). The experimental results reported in that study indicated that the proposed roadway piezoelectric energy harvester may generate enough energy from vehicles to run general devices used in the transportation infrastructure.

Recently, Khanafer and Vafai [36] numerically analyzed flow and heat transfer characteristics of a cantilever beam for piezoelectric energy harvesting for various flow speeds and thicknesses of piezoelectric layer. Their results demonstrated that the inlet velocity of the flow had a significant effect on the deflection and the electrical properties of the piezoelectric layer as depicted in Figure 5.8. Figure 5.8b shows that the maximum electric field can be generated at high flow velocity.

Several studies were conducted in the literature to investigate the performance of piezoelectric cantilever under various geometries and conditions to maximize its output power [36–41]. Goldschmidtboeing and Woias [38] conducted a theoretical study based on Rayleigh–Ritz method for piezoelectric compound structures to analyze triangular- and rectangular-shaped beams for piezoelectric energy harvesters (Figure 5.9). Their results indicated that triangular-shaped beams significantly outperformed rectangular ones in terms of tolerable excitation amplitude and maximum output power (Figure 5.10).

Benasciutti et al. [39] conducted a finite element study to analyze thoroughly the stress and strain states, as well as to evaluate the resulting voltage and charge distributions in the piezoelectric layers. Two trapezoidal configurations were used in that study (Figure 5.11). An experimental setup was constructed in that study to validate the numerical results and develop an optimized structure. Their results indicated that the reversed trapezoidal model exhibited higher power than the trapezoidal model

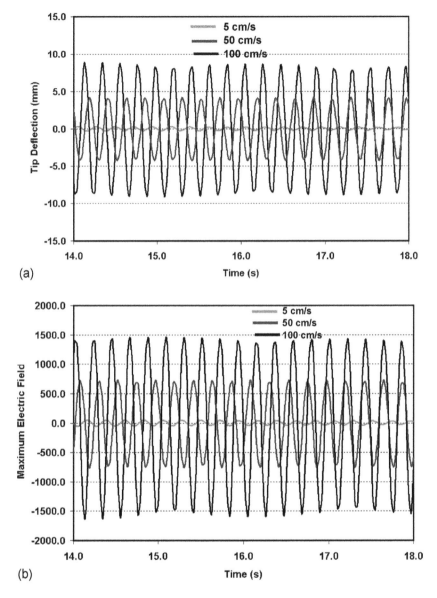

FIGURE 5.8   Effect of varying the inlet velocity on (a) the tip deflection (mm) and (b) maximum electric field (N/C). (From Jiang, X., et al., *J. Renew. Sustain. Energy*, 6, 043110, 2014. With Permission.)

of the same volume (Figure 5.12). Simon and Yves [41] conducted a numerical study to analyze the performance of different shapes of piezoelectric energy harvesters, namely a rectangular harvester, an equivalent mass/stiffness-tapered beam harvester, and an equivalent maximal strain-tapered beam harvester. Their results indicated that the tapered beam with 0.3° slope angle increased the energy harvested by 69%.

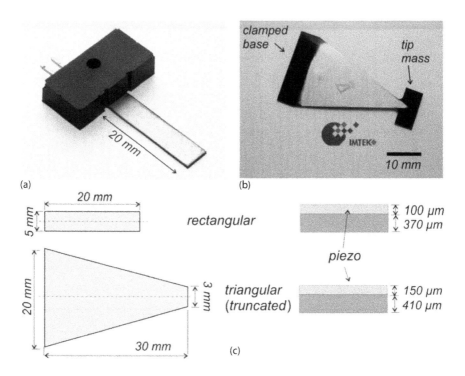

FIGURE 5.9 (a–c) Schematics of the two tested beam harvesters. (From Gong, J.J., et al., *Vibr. Test Diagn.*, 658–663, 2014. With permission.)

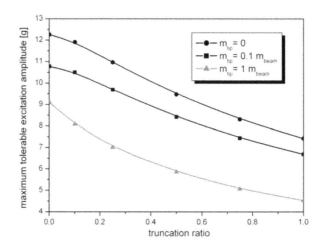

FIGURE 5.10 Maximum tolerable excitation frequency and output power for harvesters with different beam shapes. (From Gong, J.J., et al., *Vibr. Test Diagn.*, 658–663, 2014. With permission.)

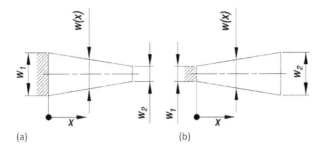

(a)   (b)

FIGURE 5.11 "Trapezoidal" (a) and "reversed trapezoidal" (b) models clamped along the side of width W1 and loaded at the free width W2. (From Goldschmidtboeing, F., and Woias, P., *J. Micromech. Microeng.*, 18, 1–7, 2008. With permission.)

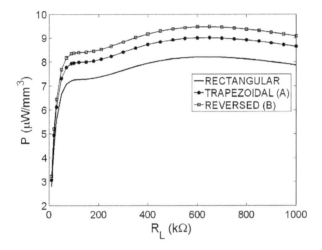

FIGURE 5.12 Effect of varying the applied resistive load on the power output for different model. (From Goldschmidtboeing, F., and Woias, P., *J. Micromech. Microeng.*, 18, 1–7, 2008. With permission.)

Roundy [42] provided a general theory that can be used to compare different approaches and designs for vibration-based generators. The author reported in that study that the trapezoidal cantilever generated twice the energy of the rectangular cantilever for a given volume. These studies proved that the geometry of the cantilever beam affects the power output. It can be noted from the literature that the performance of the piezoelectric beam depends strongly on the shape of the piezoelectric energy harvester. Thus, the sensitivity of electric field to the cantilever geometry should be examined.

The main objective of this review is to develop a finite element model in conjunction with constrained optimization-based direct search method to find out the maximum electric field and minimum structural volume while satisfying the stiffness of the piezoelectric beam. Therefore, the objectives taken into consideration are maximization of the electric field and minimization of the structural volume of the piezoelectric beam, subjected to constrains on the yield strength and width of the piezoelectric and substrate layers.

5.2 MATHEMATICAL FORMULATION

To optimize the electric field and the average stress of the piezoelectric cantilever beam while minimizing the volume of the piezoelectric beam, a finite element model was developed by using the commercial package ANSYS® in conjunction with optimization method-based direct search. The initial geometry and finite element mesh of the model used is depicted in Figure 5.13.

Grid-independence results analysis was carried out in this investigation, and it was deduced that a total number of 13,104 elements and 60,961 nodes were sufficient to produce mesh-independence results. The cantilever beam that was used in this study is composed of two layers: the top layer is a piezoelectric material and the bottom layer is an elastic material. The polarization (P) of the piezoelectric material is assumed along the y-direction. The interface between the two layers is grounded (voltage equals zero). A tip load boundary condition of 6 N was applied in the negative y-direction at the free end of the cantilever. The initial dimensions of the piezoelectric and substrate layers are listed in Table 5.1.

The mechanical and electrical properties of the piezoelectric material and the elastic substrate are given as follows [42]:

Elastic Bottom Layer: The material properties of the bottom layer are: $E = 90$ GPa, and $\nu = 0.3$

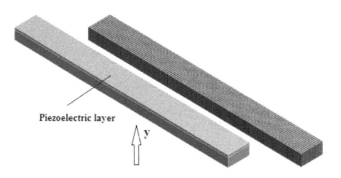

Piezoelectric layer

FIGURE 5.13   Geometry and mesh of the model used in the present investigation.

TABLE 5.1
Dimensions of the Initial Design

Parameters	Initial Value (mm)
Thickness of PZT material	1
Thickness of elastic substrate	4
Length of elastic material	100
Width of PZT material 0.5 mm $\leq L_{PZT} \leq 10$ mm	10

Piezoelectric Layer: Assuming anisotropic materials of the piezoelectric beam, the anisotropic elastic properties used in this investigation can be given in a matrix form as follows [42]:

$$C^E = \begin{bmatrix} 13.2 & 7.10 & 7.3 & 0 & 0 & 0 \\ & 13.2 & 7.3 & 0 & 0 & 0 \\ & & 11.5 & 0 & 0 & 0 \\ & & & 3 & 0 & 0 \\ & & & & 2.6 & 0 \\ & & & & & 2.6 \end{bmatrix} 10^{10} \frac{N}{m^2} \tag{5.1}$$

The piezoelectric coefficients in the strain form used in this investigation are also given in a matrix form as follows [42]:

$$e = \begin{bmatrix} 0 & -4.1 & 0 \\ 0 & 14.1 & 0 \\ 0 & -4.1 & 0 \\ 10.5 & 0 & 0 \\ 0 & 0 & 10.5 \\ 0 & 0 & 0 \end{bmatrix} \tag{5.2}$$

5.3 FORMULATION OF THE OPTIMIZATION PROBLEM

In this investigation, the optimal geometry of the piezoelectric cantilever is determined by direct search based optimization, which is used in conjunction with the ANSYS finite element software. The design problem is therefore to maximize the output electric field and simultaneously to minimize the structural volume of the piezoelectric beam, subject to the design constraints. The formulation of the optimization problem can be outlined as follows:

$$\text{Find } X = (x_1, x_2, \dots, x_k) \tag{5.3}$$

$$\text{minimize } \{V_{PZT}(X)\} \tag{5.4}$$

$$\text{and maximize } \{E_{PZT}(X) \text{ and } \sigma_{\max}(X)\} \tag{5.5}$$

$$\text{Constraints: } \{E_j \geq E_{initial,j}; V_{PZT} \leq V_{PZT,initial}; \sigma_{\max,j} \leq \sigma_y\} \tag{5.6}$$

$$\text{and } \{x_i^{\min} \leq x_i \leq x_i^{\max}, i = 1, 2, \dots, k\} \tag{5.7}$$

where X is the design variable, V_{PZT} is the volume of the piezoelectric beam to be minimized, E_{PZT} is the electric field, σ_{max} is the maximum stress (i.e., maximum Von Mises stress), and σ_y is the yield strength of 25 MPa [43]. The subscript "initial" indicates the initial value that corresponds to the initial iteration. x_i^{min} and x_i^{max} are the lower and upper limits of the design variables of x_i respectively. There was one geometrical design variable in the structural model, which represented the free end width of the piezoelectric beam. The lower and upper bounds are set to 0.5 mm and 10 mm, respectively. The optimization process starts with an initial geometry in Ansys. Each iteration, the geometry will be changed and results are obtained until maximum electric field and minimum volume are achieved.

5.4   CODE VALIDATION

The initial model is validated against analytical results reported in the literature by Smits et al. [45]. Smits et al. [45] developed a relationship between the deflection of the beam and the applied voltage using the energy density as follows:

$$\delta(x,V) = -\frac{3d_{31}Vx^2}{8t_p^2} \tag{5.8}$$

where V is the applied voltage, and t_p is the thickness of piezoelectric beam. Figure 5.14 demonstrates an excellent comparison of the deflection along the beam between the present results and the analytical results of Smits et al. [45].

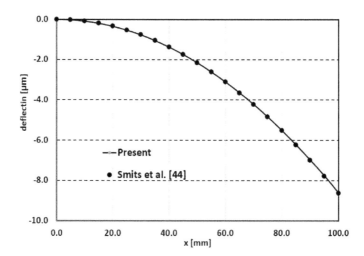

FIGURE 5.14 Comparison of the deflection along the beam between the present results and the analytical results of Smits et al. [45]. (V = 100 volt, t_p = 1 mm, d_{31} = 2.3 × 10^{-11} C/N). (From Smits, J.G., et al., *Sen. Actuators A*, 28, 781–784, 1991. With permission.)

5.5 RESULTS AND DISCUSSION

The direct optimization of the piezoelectric beam length required 20 trials to optimize the electric field of the piezoelectric beam. Figure 5.15 illustrates the optimized geometry and mesh of the model. The width of the free end was reduced to very small value of 0.5 mm compared the initial value of 10 mm, the same as at the fixed end of the beam. Table 5.2 summarizes the initial and optimized volumes of the substrate and the piezoelectric layers. This table illustrated that the structural volume of both piezoelectric and the substrate decreased considerably from 5×10^3 mm^3 to 2.62×10^3 mm^3, representing a 47.6% saving in materials.

The effect of varying the geometry on the area normalized quantities, voltage, electric field, and normal strain (ε_x) distribution along the length of the piezoelectric beam is depicted in Figure 5.16. For the studied designs, the trapezoidal beam geometry was found to produce the improved results. Figure 5.16 clearly shows that in the case of the trapezoidal geometry, the deformation increases because of the beam stiffness reduction. Consequently, the voltage, electric field, and the normal strain (ε_x) distribution along the center line of the piezoelectric beam rise significantly compared with the rectangular geometry. Figure 5.16a illustrates a comparison of the voltage/area distribution along the centerline length of the piezoelectric layer between the optimum and initial designs. The maximum voltage for both designs was found to occur in the vicinity of the fixed end due to maximum bending moment location. Moreover, optimum design was found to exhibit higher voltage/area distribution compared with the initial design as depicted in Figure 5.16a. The maximum value of

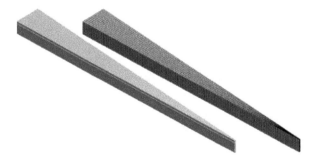

FIGURE 5.15   Optimized geometry and Finite Element Mesh.

TABLE 5.2
Volumes of the Initial and Optimized Geometries of the Substrate and Piezoelectric Layers

Parameters	Initial Geometry (mm³)	Optimized Geometry (mm³)
Volume of the substrate	4×10^3	2.09×10^3
Volume of the piezoelectric	1×10^3	5.23×10^4
Total volume	5×10^3	2.62×10^3

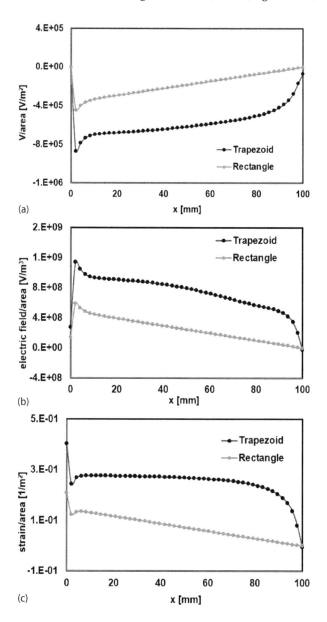

(a)

(b)

(c)

FIGURE 5.16 Comparison of the area normalized quantities (a) voltage, (b) electric field, and (c) normal strain (ε_x) distribution along the center length of the piezoelectric beam between the rectangular design and the trapezoidal design for the same tip force.

the voltage/area was found to increase by 96.4%. Figure 5.16b demonstrates that the maximum electric field per unit area was also found to increase from 5.87×10^8 V/m³ for the initial design to 1.14×10^9 V/m³ for the optimum design, which represents an increase of 94.2%. As can be seen from Figure 5.16c, the trapezoidal shape exhibits higher strain than the rectangular shape for identical load.

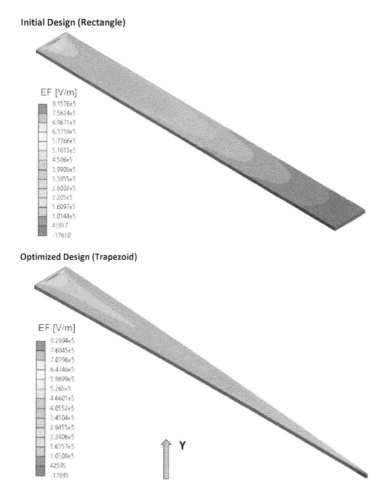

FIGURE 5.17 Comparison of the electric field distribution in y-direction between the rectangular design and the trapezoidal design of the piezoelectric beam (tip force of the beam = 6 N).

Figure 5.17 shows the contour plots for the comparison of the electric field distribution in y-direction between the initial rectangular design and the optimized geometry (trapezoidal design) of the piezoelectric beam (tip force of the beam = 6 N). The electric field is found maximum in the vicinity of the fixed end for both designs, and this is associated with maximum bending moment location. Furthermore, the trapezoidal design shows a more even distribution than the rectangular design.

A similar behavior is also observed, when the voltage distribution is considered, as depicted in Figure 5.18. The von Mises stress fields in both designs are compared, under the same tip load of 6 N, applied downward as shown in Figure 5.19. Although the structural volume of the optimized trapezoidal design reduced by 47.6% from its initial volume, the maximum values of the induced stresses in both designs are approximately of the same order (about 17 MPa), and they occur at the constrained

Initial Design (Rectangle)

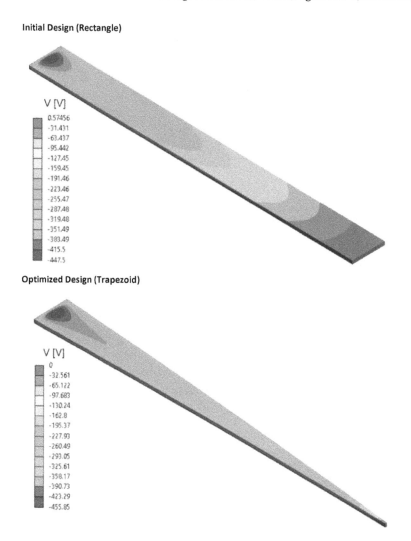

Optimized Design (Trapezoid)

FIGURE 5.18   Comparison of the voltage distribution between the rectangular design and the trapezoidal design of the piezoelectric beam (tip force of the beam = 6 N).

end of each beam. The stress distribution in the trapezoidal beam shows less variation along the beam axis. The minimum stress value occurs in both cases at the free beam end, whereas a much higher value arises at the tip of the trapezoidal beam due to the very small cross-sectional area (0.25 MPa vs. 0.02 MPa).

Modal analysis was carried out on both beam designs to calculate the first four natural frequencies, given in Table 5.3. Figure 5.20 displays the resulting frequencies compared with the number of vibration mode for both beams. At all calculated modes, the trapezoidal cantilever design has higher natural frequency values than the rectangular design. The increased sensitivity of the trapezoidal design leads to higher power output.

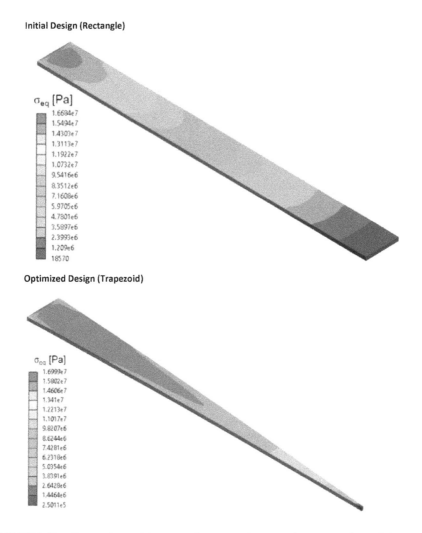

FIGURE 5.19   Comparison of the von Mises stress between the rectangular and the trapezoidal designs.

TABLE 5.3
The First Four Natural Frequencies for Rectangular and Trapezoidal Cantilever Beam Designs

Mode	Rectangular Geometry f[Hz]	Trapezoidal Geometry f[Hz]
1	331.07	619.12
2	2046.7	2669.6
3	5594.8	6388.5
4	9830.1	11509

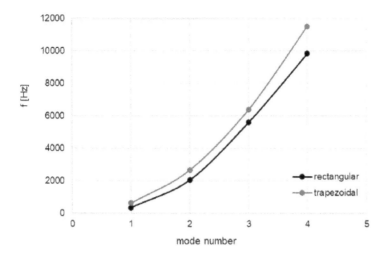

FIGURE 5.20 Comparison of the natural frequency between the rectangular design and the trapezoidal design for different modes.

5.6 CONCLUSIONS

Finite element method in conjunction with direct optimization was conducted in this review to maximize the output electric field of the piezoelectric beam while minimizing the structural volume. The present numerical method was validated against theoretical studies reported in the literature and excellent agreement was found between both results. The results of this review indicated that the structural volume of both piezoelectric and the substrate reduced significantly from 5×10^3 mm^3 to 2.62×10^3 mm^3, representing a 47.6% saving in materials. The maximum Von Mises stress results showed insignificant increase from 16.8 MPa to 17 MPa, respectively. These results are well within the strength constrains ($\sigma_{yield} = 24$ MPa). The maximum electric field per unit area was found to increase from 5.87×10^8 V/m^3 for the initial design to 1.14×10^9 V/m^3 for the optimum design, which represents an increase of 94.2%. The modal analysis illustrated that the optimum design (trapezoidal cantilever) exhibited higher natural frequencies at all calculated modes, which may lead to higher power output. The preliminary results presented in this review revealed that varying the geometry of the piezoelectric layer may have a significant effect on the characteristics of piezoelectric beam. Therefore, the future work will focus on maximizing the output power by seeking optimum topology of the cantilever beam.

REFERENCES

1. Sonar HA, Paik J. Soft pneumatic actuator skin with piezoelectric sensors for vibrotactile feedback. *Frontiers in Robotics and AI.* 2016;2(38):1–11.
2. Cook-Chennault K, Thambi N, Sastry AM. Powering MEMS portable devices– A review of non-regenerative and regenerative power supply systems with special emphasis on piezoelectric energy harvesting systems. *Smart Materials and Structures.* 2008;17:043001.

3. Ghosh SK, Mandal D. Bio-assembled, piezoelectric prawn shell made self-powered wearable sensor for noninvasive physiological signal monitoring. *Applied Physics Letters.* 2017;110:123701.

4. Liu Z, Zhang S, Jin YM, Ouyang H, Zou Y, Wang XX, et al. Flexible piezoelectric nanogenerator in wearable self-powered active sensor for respiration and healthcare monitoring. *Semiconductor Science and Technology.* 2017;32:064004.

5. Ghosh SK, Adhikary P, Jana S, Biswas A, Sencadas V, Gupta SD, et al. Electrospun gelatin nanofiber based self-powered bio-e-skin for health care monitoring. *Nano Energy.* 2017;36:166–175.

6. Lu C, Wu S, Lu B, Zhang Y, Du Y, Feng X. Ultrathin flexible piezoelectric sensors for monitoring eye fatigue. *Journal of Micromechanics and Microengineering.* 2018;28:025010.

7. Curry EJ, Ke K, Chorsi M, Worbel KS, Miller AN, Patel A, et al. Biodegradable piezoelectric force sensor. *Proceedings of the National Academy of Sciences.* 2018;115(5):909–914.

8. Markus DT, Hayes MC, inventors. Piezoelectric sensor for vision correction. US patent 20160030160A1. April 25, 2017.

9. Chiu Y, Lin W, Wang H, Huang SB, Wu M. Development of a piezoelectric polyvinylidene fluoride (PVDF) polymer-based sensor patch for simultaneous heartbeat and respiration monitoring. *Sensors and Actuators, A.* 2013;189:328–334.

10. Sonar HA, Paik J. Soft pneumatic actuator skin with piezoelectric sensors for vibrotactile feedback. *Frontiers in Robotics and AI.* 2016;2(38):1–11.

11. Khanafer K, Vafai K. Analysis of heat transfer and flow characteristics of a microcantilever beam for piezoelectric energy harvesting. *International Communications in Heat and Mass Transfer.* 2018;98:265–272.

12. Khanafer K, Vafai K, Gaith M. Fluid–structure interaction analysis of flow and heat transfer characteristics around a flexible microcantilever in a fluidic cell. *International Communications in Heat and Mass Transfer.* 2016;75:315–322.

13. Khaled ARA, Vafai K. Analysis of detection enhancement using microcantilevers with long-slit-based sensors. *Sensors.* 2013;13(1):681–702.

14. Khanafer K, Vafai K. Geometrical and flow configurations for enhanced microcantilever detection within a fluidic cell. *International Communications in Heat and Mass Transfer.* 2005;48(14):2886–2895.

15. Khanafer K, Khaled ARA, Vafai K. Spatial optimization of an array of aligned microcantilever based sensors. *Journal of Micromechanics and Microengineering.* 2004;14(10):132.

16. Khaled ARA, Vafai K, Yang M, Zhang X, Ozkan CS. Analysis, control and augmentation of microcantilever deflections in bio-sensing systems. *Sensors and Actuators B: Chemical.* 2003;94(1):103–115.

17. Park G, Rosing T, Todd MD, Farrar CR, Hodgkiss W. Energy harvesting for structural health monitoring sensor networks. *Journal of Infrastructure Systems.* 2008;14(1):64–79.

18. Garcia LR, Lunadei L, Barreiro P, Robla JI. A review of wireless sensor technologies and applications in agriculture and food industry: State of the art and current trends. *Sensors.* 2009;9(6):4728–4750.

19. Ko J, Lu C, Srivastava MB, Stankovic JA, Terzis A, Welsh M. Wireless sensor networks for healthcare. *Proceedings of the IEEE.* 2010;98(11):1947–1960.

20. Liao WH, Wang DH, Huang SL. Wireless monitoring of cable tension of cable-stayed bridges using PVDF piezoelectric films. *Journal of Intelligent Material Systems and Structures.* 2001;12(5):331–339.

21. Chalard L, Helal D, Verbaere L, Wellig A, Zory J. Wireless sensor networks devices: Overview, issues, state-of-the-art and promising technologies. *ST Journal of Research.* 2007;4(1):4–8.
22. Carlos FG, Pablo HI, Joaquin GH, Jesus AP. Wireless sensor networks and applications: A survey. *International Journal of Computer Science and Network Security.* 2007;7(3):264–273.
23. Akyildiz I, Su W, Sankarasubramaniam Y, Cayirci E. Wireless sensor networks: A survey. *Computer Networks.* 2002;38(4):393–422.
24. Raghunathan V, Schurgers C, Park S, Srivastava MB. Energy-aware wireless microsensor networks. *IEEE Signal Processing Magazine.* 2002;19(2):40–50.
25. Roundy S, Wright PK, Rabaey J. A study of low level vibrations as a power source for wireless sensor nodes. *Computer Communications.* 2003;26(11):1131–1144.
26. Jiang X, Polastre J, Culler D. Perpetual environmentally powered sensor networks. In: *Proceedings of the 4th International Symposium on Information Processing in Sensor Networks (IPSN'05),* pp. 463–468, April 2005.
27. Hausler E, Stein E. Implantable physiological power supply with PVDF film. *Ferroelectrics* 1984;60:277–282.
28. Umeda M, Nakamura K, Ueha S. Analysis of the transformation of mechanical impact energy to electric energy using piezoelectric vibrator. *Japanese Journal of Applied Physics.* 1996;35:3267–3273.
29. Ramsey MJ, Clark WW. Piezoelectric energy harvesting for bio MEMS applications. In: *Proceedings of SPIE's 8th Annual Smart Materials and Structures Conference,* Vol. 4332, Newport Beach, CA, pp. 429–438, 2001.
30. Kymissis J, Kendall C, Paradiso J, Gershenfeld N. Parasitic power harvesting in shoes. In: Second International Symposium on Wearable Computers, pp. 132–139, 1998.
31. Starner T. Human-powered wearable computing. *IBM Systems Journal.* 1996; 35(3–4):618–629.
32. Qian F, Xu T-B, Zuo L. Design, optimization, modeling and testing of a piezoelectric footwear energy harvester. *Energy Conversion and Management.* 2018;171:1352–1364.
33. Hwang GT, Park H, Lee JH, Oh S, Park KI, Byun M, et al. Self-powered cardiac pacemaker enabled by flexible single crystalline PMN-PT piezoelectric energy harvester. *Advanced Materials.* 2014;23:4880–4887.
34. Yang H, Wang L, Zhou B, Wei Y, Zhao Q. A preliminary study on the highway piezoelectric power supply system. *International Journal of Pavement Research and Technology.* 2018;11:168–175.
35. Jiang X, Li Y, Li J, Wang J, Yao J. Piezoelectric energy harvesting from traffic-induced pavement vibrations. *Journal of Renewable and Sustainable Energy* 2014;6:043110.
36. Khanafer K, Vafai K. Analysis of heat transfer and flow characteristics of a microcantilever beam for piezoelectric energy harvesting. *International Communications in Heat and Mass Transfer.* 2018;98:265–272.
37. Gong JJ, Xu YY, Ruan ZL, Bian YX. Simulation on generating capacity for energy harvesting device with piezoelectric bimorph cantilever. *Journal of Vibration, Measurement and Diagnosis.* 2014;14:658–663.
38. Goldschmidtboeing F, Woias P. Characterization of different beam shapes for piezoelectric energy harvesting. *Journal of Micromechanics and Microengineering.* 2008;18(10):1–7.
39. Benasciutti D, Moro L, Zelenika S. Vibration energy scavenging via piezoelectric bimorphs of optimized shapes. *Microsystem Technologies.* 2010;16(5):657–668.
40. Dietl JM, Garcia E. Beam shape optimization for power harvesting. *Journal of Intelligent Material Systems and Structures.* 2010;21(6):633–646.

41. Simon P, Yves SA. Electromechanical Performances of Different Shapes of Piezoelectric Energy Harvesters. International Workshop Smart Materials and Structures, Montreal, Canada, October 22–23, 2009.

42. Roundy S. On the effectiveness of vibration-based energy harvesting. *Journal of Intelligent Material Systems Structures.* 2005;16:809–823.

43. Lange R, Scampoli S, Ansari N, Shaw D. Piezoelectric Fan Modeling FSI Analysis using ANSYS and CFX, International ANSYS Conference, Pittsburgh, USA, 2008.

44. Bert CW, Birman V. Effects of stress and electric field on the coefficients of piezoelectric materials: One-dimensional formulation. *Mechanics Research Communications.* 1998;25(2):165–169.

45. Smits JG, Dalke SI, Cooney TK. The constituent equations of piezoelectric bimorphs. *Sensors and Actuators A.* 1991;28:781–784.

Cody S. Long and Ahmad Vasel-Be-Hagh

CONTENTS

6.1 INTRODUCTION

As far back as history has been recorded, the use of renewable energy has been seen for different applications rather than generating electrical power. Sørensen [1] writes that wood for fire fuel dated back 10,000 years, wind energy was used 5,500 years ago for sailboats, and solar power was used to ignite fires. As technology and society advanced, people became more reliant on fossil fuels as the need for more immediate and reliable power exploded. In today's world, however, the need for renewable energy is growing every day for multiple reasons, including energy security [2–4], environmental issues [5–7], and the fact that fossil fuels are finite [8,9]. Dependency on fossil fuels can pose national security risks because oil-dependent countries often require military involvement to ensure there are no disruptions in the oil supply, while oil suppliers weaponize the energy market in pursuit of scoring other political points [2,3]. According to Krane [5], the burning of fossil fuels accounts for two-thirds of the world's greenhouse gas emissions. This alone should be a strong motivating reason to push toward a more sustainable energy industry. Shafiee and Topal [8] stated that oil and gas reserves, based on the current ratio of reserve to use, will both run out in less than 100 years. With fossil fuel sources and reserves being so finite, utilization of naturally occurring energy sources are the way to survive in the future. Because the need for renewable energy sources was discovered, extensive research has been done on these technologies. The engineering community is making leaps and bounds in this area of research, but the need to do more is still present. As more research is conducted, major issues such as unpredictability and intermittence (e.g., wind not blowing, water flow dropping, cloudy sky), as well as low-conversion efficiency are flagged, making raw renewable sources non-dispatchable to the grid. Hence, using renewable sources in a raw manor (i.e., using them directly as reliable methods of power production is not possible). One viable option to overcome the intermittency and unpredictability issues to make renewable energy mainstream and scalable to the commercial level is developing energy storage plants. Although the technology for storing energy exists, it is not yet affordable for a full transition from intermittency to dispatchability. To reduce expenses and technical complications and increase the roundtrip efficiency, there is a great interest in incorporating integrated storage techniques so that they share parts with the renewable energy harvester. Mixing the storage and harvester, rather than having two separate storage and power-generation plants, would also save space and would facilitate the process of obtaining required permits for building new power plants. In this chapter, a review of existing integrated storage methods will be discussed. Within each subsection, examples of an application for each will be presented. Between each of these examples, efficiencies will be compared and discussed as applicable. From these findings, advantages and disadvantages will be given for each. A new technology, termed SAVER (Storage-integrAted Vortex hydrokinetic Energy converteR), will be introduced and compared against other existing generation-storage technologies. At the heart of SAVER, there is a flexible cylinder that will use the natural flow of bodies of water such as oceans, seas, and rivers to generate electricity via dynamics of vortex-induced vibrations, while also serving as an integrated storage module. The surplus electrical power of the grid is saved in the form of potential energy of compressed air accumulated in those flexible cylinders. When desired, the compressed air stored in the cylinders would be released back into the free surface to run some turboexpanders to regenerate electricity. Upon completion of reviewing each individual integrated storage technology, a collective comparison is presented.

6.2 INTEGRATED STORAGE METHODS

So many energy storage technologies have been proposed and developed in industrial scale, not all of them, however, have the flexibility to be integrated within renewable energy-generation technologies. In this section, the storage technologies that have been incorporated into renewable energy harvesters are introduced and discussed. These technologies include hydro pumped storage, flenerat, super capacitors, batteries, buoyancy-based storage, and compressed air energy storage.

6.2.1 HYDRO PUMPED STORAGE (HPS)

Hydro pumped storage (HPS) stores electrical energy in the form of gravitational potential energy by pumping water to areas of higher elevations during hours of low power demand, then released during hours of high demand. When released, the potential energy of the water is converted into kinetic energy and then back into electrical power when water runs through a turbine. Traditionally, the biggest downside to using HPS integrated systems was location dependency. It can only be used in locations where there is a considerable difference in elevation between the water source and the storage area. Other issues arise when considering environmental, aesthetic, and land impact [10]. The best use of this technology is in parallel with other energy-generation technologies. For example, integrating HPS with wind energy can be an effective combination because the HPS can be a quick response solution to intermittency in wind power generation [11,12].

In a brand-new concept by Pali and Vadhera [11], it is proposed to use open wells located in rural areas to generate stable and predictable power with minimized intermittence by combining wind turbine and HPS technologies. The HPS system operates using the initial power generated by the wind turbine to fill up an upper reservoir. After running through a control valve, the water stored in the reservoir will steadily turn a turbine and charge a generator [11]. This generator provides the stable system output (Figure 6.1). The beauty of this system is that it is capable of operating

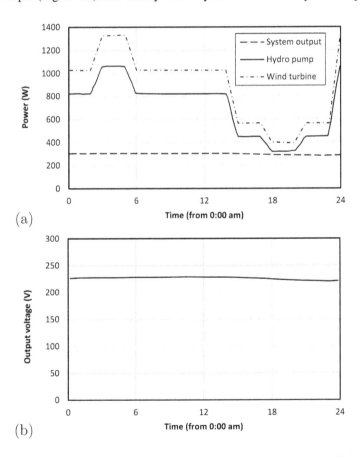

FIGURE 6.1 (a and b) Output of the proposed hydro pumped storage (HPS)-integrated wind turbine using open wells. (From Pali, B., and Vadhera, S., *Renew. Energy*, 127, 802–810, 2018.)

for a period of time even if there is no wind power. This period of operation without any wind power was found to be approximately 4 hours for the case studied by Pali and Vadhera, assuming the upper reservoir is full [11]. Although a system efficiency was not given, plots were provided that showed even through variable wind turbine power, there was an almost constant system output of power and voltage that was sufficient to provide power to the proposed community (Figure 6.1) [11]. This research provides data that could, in the future, change the way that HPS systems are used.

HPS can also be useful in areas with large rainfall or flooding, as stated in the work of Safi and Basrawi [12]. In this research, the authors discuss the effects, both environmentally and economically, of integrating a reservoir storage system with a wind-diesel power generation system compared to using a stand-alone wind-diesel power generation system and a stand-alone diesel power system. When comparing the three system configurations, the authors used the Hybrid Optimization Model for Electric Renewable (HOMER) software to simulate the actual conditions. Although the research, as reported in their article, does not lead to a specific efficiency for each system, the results provided speak on the value of the system in terms of increasing penetration of wind power (see Figure 6.2) [12]. It can be easily seen how much more efficient the system with the flood mitigation is. While the flood mitigation system produced the least total amount of electricity (in kWh/yr), the electricity produced was nearly 100% from the wind turbine, indicating the significant impact of integrated pumped HPS on increasing penetration of wind from approximately 12% to almost 100%.

When considering the aforementioned HPS integrated systems, there is a common theme: The best use of hydro-energy storage, both economically and technologically, is in conjunction with a renewable energy harvester. The problem that seems to be prevalent in literature is the fact that at the present time, integrated

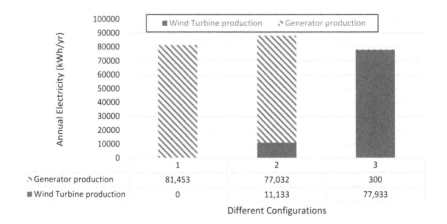

FIGURE 6.2 Hydro pumped storage (HPS)-integrated wind turbine using flood water. Comparing electricity generation by different configurations: (i) diesel generator only, (ii) diesel generator + wind turbine, and (iii) diesel generator + wind turbine + storage reservoir. (From Safi, A., and Basrawi, F., *IOP Conference Series: Materials Science and Engineering*, 2018.)

hydro energy storage systems do not have a very large hold in the consumer market. Although it can be a reach to directly compare these systems, as one of them is for a macro-scale use [12] while the other one is proposed for smaller scales [11], comparisons can be drawn when looking at the overall savings that can come from such integration. The research by Pali and Vadhera involves using the natural resources of the studied area (wind and well water) to generate power to run a house [11]. Even with wind being a common resource, it is stated that with a completely filled upper reservoir, the system could operate for nearly 4 hours with no wind [11]. One could readily find the wind patterns for an area along with the price of electricity and see the savings. The research by Safi and Basrawi [12] shows that a system with hydro storage can nearly eliminate the use of a generator when used in tandem with a wind turbine. This system was used in macro scale for flood mitigation. Looking at both small- and large-scale concepts, one can see how there is a financial benefit of using an integrated hydro energy storage system if the initial cost of implementing the system can be justified feasible.

6.2.2 FLYWHEEL

A flywheel is a device that stores kinetic energy by accelerating a rotor with high moment of inertia to very high speeds. It maintains the energy in the form of rotational kinetic energy. When desired, the flywheel is engaged to the generator to convert back the stored energy at rates beyond the ability of the original energy source [10]. Flywheels are useful because they do not require very much space to operate and have fairly low maintenance. However, due to the capital cost, flywheels are mainly used in commercial and macro-scale systems. In a research conducted by Prodromidis and Coutelieris [13], simulations were carried out to test the economic and technical feasibility of using flywheel-integrated renewable energy systems. The purpose of the research is to assess if flywheels could be integrated into everyday renewable energy systems. It should be noted that the simulations presented are based on the load for a typical house located on Naxos island due to its high renewable energy sources; the system is assumed to be used simultaneously with solar and wind energy; and finally, a stack of batteries was used to support the flywheel [13]. The authors propose hooking a flywheel to a DC-AC converter and, in turn, a battery [13]. To achieve the most reliable results, the authors developed six different scenarios to test (Table 6.1). The operational characteristics are presented in Table 6.2.

The most noticeable bit of information available from Table 6.1 is the reduction of batteries needed for the setups involving the flywheel compared with their original setup without the flywheel. Coming directly from this information, in Table 6.2 one can see how efficient each flywheel and battery combination will be. There are a lot of factors that can be taken from this table, with arguably the most important being the roundtrip efficiency. Being able to maintain more than 75% of the stored energy is an impressive feat. It should also be noted how all of the combinations have a long lifetime. A comparison of costs and the future of flywheel storage systems is also conducted. The authors write "it is observed that although the initial cost of the systems with simple batteries are much lower, systems combining flywheels can be competitive because the NPC (Net Present Cost) of the different systems are

TABLE 6.1
List of Scenarios Simulated by Prodromidis and Coutelieris to Evaluate Effectiveness of Integrating Flywheel into Wind/Solar Systems for a Desirable Load of 7,775 kWh

Systems	PV (kW)	Wind (kW)	Batteries	New Battery Configuration
PV, wind, and Hoppecke batteries	1	1 × 5	4 (Hoppecke 3,000 Ah)	*
PV, wind, and Surrette batteries	1	1 × 5	6 (Surrette 1,900 Ah)	*
PV, wind, and vision batteries	1	1 × 5	58 (Vision 55 Ah)	*
PV, wind with flywheel, and Hoppecke batteries	1	1 × 5	*	2 (Hoppecke 3,000 Ah)
PV, wind with Flywheel, and Surrette batteries	1	1 × 5	*	5 (Surrette 1,900 Ah)
PV, wind with flywheel, and vision batteries	1	1 × 5	*	10 (Vision 55 Ah)

TABLE 6.2
Characteristics of the Scenarios Presented in Table 6.1

Operational Characteristics	Flywheel + Hoppecke	Flywheel + Surrette	Flywheel + Vision
Roundtrip efficiency %	75.2	75.2	75.2
Nominal capacity (Ah)	4,250	2,525	263.33
Nominal voltage (V)	2	4	12
Minimal state of charge %	30	40	40
Float life years	30	30	30
Max charge rate (A/Ah)	3.926	3.856	3.606
Max charge current (A)	610	67.5	16.5
Lifetime throughput (Cycles)	150,000	150,000	150,000

Source: Prodromidis, G.N. and Coutelieris, F.A., *Renew. Energy*, 39, 149–153, 2012.

equivalent. … Finally, this innovative method of energy storage shows that flywheel systems could be commercialized in the near future" [13]. From the information, it can be seen that there are several advantages to using a flywheel-integrated renewable energy system. By adding the flywheel, the total number of batteries needed dropped in each setup (Table 6.1). The flywheel and battery tandems proposed in this concept yield some of the highest efficiency values found in the complete review of the present chapter. This combination of factors shows that in the right situation, a flywheel-integrated system could be an efficient method of energy storage. It should also be noted that the research by Prodromidis and Coutelieris was done in a specific location that had an abundance of the needed natural resources [13]. More work should be presented to find the effect of location. Some of the disadvantages of using a flywheel are the safety issues that arise. When running, a flywheel has rapidly rotating parts. This could be fatal if a part was pulled into the wheel, or if failure were to occur. Another issue that can arise is that flywheels are limited by their size. This could cause issues when trying to design a stand-alone flywheel system.

6.2.3 SUPER CAPACITORS

Super capacitors (SC) are electromechanical capacitors with very high-power density (i.e., ability to release energy at very high rates [14]), while their energy density is low. SCs are similar to batteries but serve the opposite purpose. Batteries have a high energy density; however, their discharge rate is significantly smaller than SCs. SCs are also useful for their life cycle because they can go through many cycles before degradation occurs. The main issue with SCs is being very costly relative to other technologies. Considering the low energy density of each SC, it is easy to notice how rapidly the implementation costs would increase if they are used in conjunction with renewable energy harvesters, for instance, a wind farm. Based on the relevant knowledge, SCs would be most useful in small-scale systems where many are not needed, such as biomedical sensors powered via piezoelectric materials.

6.2.4 BATTERIES

Batteries are devices that use a chemical process to generate electricity. High density is one of the most outstanding characteristics of batteries. Batteries also have a large life span. They can be loaded and unloaded over a large number of cycles. Although batteries offer major advantages, there are also several negative aspects associated with them. Batteries can be affected by temperature. This limits the applications and locations they can be used. Batteries are also not economical because they can be expensive to purchase in large quantities for macro-scale plants and to dispose of. The disposal issue is not only costly but also bad for the environment. So, although batteries can be useful for certain applications, the overall impact must be addressed before a decision is made.

Although it is easy for one to simply consider the negative qualities of batteries and assume there are much better options to consider for energy storage than them, batteries remain the most used storage methods when integrated with other forms of renewable energy-generation devices [15]. Nearly every research, if not all, that was reviewed for the present chapter proposes a new or improved method of alternative energy storage that used batteries as the main or backup storage medium.

According to Tiwari et al. [15], battery storage offers outstanding techno-economic advantages leading to dominance among all available energy storage technologies. Tiwari et al. [15] compared batteries to other energy-storage systems on the premise that the draining of conventional fuels and climate change has warranted a need for the reliance on renewable energy for main stream power-generation systems [15]. It is then proposed to integrate a wind-battery system with thermal generators, which are actually capable of meeting the intermittency demand alone [15]. According to their economic analysis, substituting the thermal generator system with the integrated thermal-wind-battery system leads to daily savings of nearly $66,600 [15]. The detailed cost data for the thermal generator system and the integrated thermal wind-battery system are presented in Tables 6.3 and 6.4, respectively. Although the efficiency values are not provided, the improvement of using a battery-integrated system can be seen in the economic saving shown in these tables.

The validity of using batteries in conjunction with a renewable energy-generation configuration is further confirmed by a research conducted at the University of Nottingham [16]. This study demonstrates the necessity of having better control of the amount of power generated from renewable-based power generators to ensure

TABLE 6.3
Cost for the Generator Only System

Hours of Operation	Fuel Cost $	Start-up Cost $
1	13,683	0
2	14,554	0
3	16,302	0
4	18,598	900
5	20,020	560
6	21,860	0
7	23,262	1,100
8	24,150	0
9	26,589	340
10	29,366	520
11	31,220	60
12	33,205	60
13	29,366	0
14	26,589	0
15	24,318	0
16	20,896	0
17	20,020	0
18	21,860	0
19	24,318	170
20	30,164	670
21	26,589	0
22	21,879	0
23	17,795	0
24	16,053	0
	Total Fuel $ 552,656	Total Start-up $ 4,380
	Total $ 557,036	

Source: Tiwari, S., et al., *Int. J. Renew. Energy Res.*, 8, 692–701, 2018.

TABLE 6.4
Cost for Thermal Generation with BES-Integrated WES

Hours of Operation	Fuel Cost $	Start-up Cost $
1	13,683	0
2	14,554	0
3	15,847	0
4	18,598	900
5	20,020	560
6	21,860	0
7	23,262	1,100
8	24,150	0
9	26,179	340
10	28,979	520
11	31,220	60
12	33,205	60
13	29,366	0
14	22,367	0
16	18,237	0
17	15,362	0
18	14,137	0
19	18,603	170
20	23,718	670
21	20,468	0
22	16,678	0
23	12,168	0
24	8,039	0
	Total Fuel $ 486,113	Total Start-up $ 4,380
	Total $ 490,493	

Source: Tiwari, S., et al., *Int. J. Renew. Energy Res.*, 8, 692–701, 2018.

the demand is met, rather than having a nonstop production at maximum amount of available power, which is actually intermittent. This requires storage-integrated harvesters to store energy and then use it to overcome any fluctuations that arise from intermittency. Such fluctuations cause overcharging and fines that are occurring from both consumers and wind farm operators. Batteries are stated as "perhaps the most versatile than any of the storage devices [to overcome the intermittency and unpredictability issues from renewables] as they offer desirable characteristics for wide ranges of applications and are generally cheaper in most cases" [16]. Batteries are arguably the most developed of all energy storage systems. Although there are so many implementations that batteries can be used in, they can be expensive to purchase and set up, and disposal is horrible for the environment. The pros and cons should definitely be weighed before using a battery-integrated storage system.

6.2.5   BUOYANCY-BASED ENERGY STORAGE

The offshore buoyancy-based storage, proposed by Bassett et al. [17], uses surplus electrical power to run a motor-driven pulley system to pull down a buoyant object through an open body of water. When desired, the object will be released to travel back to the surface under the buoyancy force. The linear motion of the object will then be converted into a rotary motion to drive a generator and recreate the required electrical power. One of the most attractive aspects of this system is the scalability factor. Because of the systems setup, the primary elements needed for increased storage capacity are air and water, unlike chemical-based batteries [17]. The proposed concept has the potential storage capabilities on the magnitude of Gigawatt-hour, which is a level currently available with the use of hydro pumped and compressed air energy storage [17]. The largest stated issue is the cost of the system because it needs expensive offshore below-surface constructions requiring divers or robotic-operated vehicles. Such constructions are several times more expensive than terrestrial constructions. When developing this storage system, one of the important aspects to consider was the hydrodynamic drag [17]. As seen in Table 6.5, different drag losses are listed along with the efficiency ranges. The authors next calculated the roundtrip efficiency, which is arguably the most important aspect of the entire concept. By taking into account the efficiency of motor, generator, charge, discharge, and pulley, the roundtrip efficiency can be calculated as:

$$\eta_{roundtrip} = \eta_{motor}\ \eta_{generator}\ \eta_{charge}\ \eta_{discharge}\ \eta_{pulley} \tag{6.1}$$

All the efficiency values are standardly used or provided. The efficiency of the system can be estimated as 83%, since: $\eta_{roundtrip} = 0.97 \times 0.95 \times 0.97 \times 0.97 \times 0.96 = 0.83$.

6.2.6   COMPRESSED AIR ENERGY STORAGE

Compressed air energy storage (CAES), in two different forms, underground and underwater, is a technology that has been used for decades. In CAES plants, energy is stored in the form of potential energy of compressed air that is accumulated in

TABLE 6.5
Hydrodynamic Losses for Considered Floats

Float Model	Max Drag Loss (Wh)	Drag Loss @ Crisis (Wh)	U @ Drag Crisis (m/s)	Efficiency Range (%)
5T	35.28	0.58	0.2	97.88–98.96
10T	20.46	0.44	0.151	99.10–99.98
20T	32.87	0.54	0.13	99.41–99.99
35T	37.8	0.63	0.115	99.59–99.99

Source: Bassett, K., et al., *J. Energy Storage*, 14, 256–263, 2017.

either rigid, underground tanks or flexible, underwater balloons. CAES units operate like turbines except that the compression and expansion cycles occur at different times [11]. CAES has critical advantages compared to other storage technologies. They are relatively low-cost systems that can store large amounts of energy for medium to large scale power [11,18,19]. When compared to other forms of energy storage, CAES is generally cheaper than most, size for size [20]. CAES can be used in a large number of places. It is scalable and environmentally friendly. It can be used both inland and offshore. The biggest issue that is associated with CAES is the overall low efficiency that comes from the losses that occur during the conversion processes (roundtrip efficiency of approximately 60%–70% [21]). It is generally known that losses would occur within any type of conversion, storage, and heat transfer cycle, so it should come as no surprise that the efficiency is relatively low. Like all other storage systems discussed in this chapter, CAES technology can be mixed with alternative energy harvesting technologies. So far, two concepts have been proposed to mix CAES into (i) solar energy and (ii) vortex hydrokinetic energy harvesters to store excess energy during low demand times to regulate the power output.

6.2.6.1 CAES-Integrated Photovoltaic Plant

The concept of floating storage-integrated photovoltaic (PV) plants using underwater CAES is proposed by Cazzaniga et al. [19]. As shown by Cazzaniga et al. [19], efficiency of CAES increases to approximately 80% when it is mixed into floating PV panels. The floating aspect of this concept alleviates some negative impacts of the PV plants, including the need for large land areas, which are driven by the low power density of solar energy. In the literature, there are two different systems that are proposed for floating PV arrays: (i) a raft structure that will support the PV panels with a pontoon system on the surface of the water and (ii) a submerged system that uses floats to hold the PV panel at a certain depth of water while it is anchored onto the bottom of the body of water [19]. For the storage module, underwater compressed air storage is preferred over well-developed batteries because the batteries are costly with a shorter lifetime and cause environmental and disposal difficulties. Underwater CAES, however, is environmentally friendly, inexpensive, and scalable. The proposed design was taken a step further by making the compressed air reservoir an "integral part" of the platform [19]. Pontoons of the support structure were used as the accumulators of high-pressure air to reduce costs and required space. In this technology, air is compressed through an isothermal process so the heat created through the compression phase can be efficiently transferred to an external thermal bath. The isothermal process is facilitated by reducing the speed of the process and providing large heat transfer interfaces. Another version of this technology is also presented where the compression occurs through an adiabatic process, using large insulated, high-pressure reservoirs. The adiabatic version of this technology does not require a thermal energy storage [19]. Mixed approaches that combine features from both adiabatic and isothermal compression have been used, too.

Working principles of the isothermal version of this technology are: (i) the floating PV platform will supply electric energy through modules and inverters to the grid or air compressor, (ii) the air compressor will increase the pressure in the

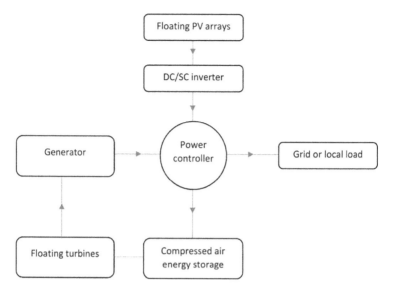

FIGURE 6.3 Block flow scheme for compressed air energy storage (CAES)-integrated floating photovoltaic (PV) plants. (From Cazzaniga, R., *J. Energy Storage*, 13, 48–57, 2017.)

pontoons from 0.1 to 20 MPa, and (iii) the compressed air can be used whenever necessary to produce energy through a turbine. The heat stored in the thermal bath will be added to the compressed air before it enters the turbine during the discharge process. The flow scheme for the system is given in Figure 6.3. This scheme shows how the system takes in energy from the PV array while having access to the energy from the compressed air storage module. The direction of power flow is all controlled by a power controller. The amount of energy stored in the pontoons is calculated as:

$$E_{Stor} = n\,Rx\,T_{(ca,f)}\,ln\frac{P_{p,f}}{P_{p,i}} \qquad (6.2)$$

where R is the gas constant, $P_{p,i}$ and $P_{p,f}$ are the initial and final air pressure at the pontoons, and T_{ca} is the temperature of the compressed air. Equation (6.2) was used to identify the optimal range of pressure for the pontoons. Since the description is assumed for a perfectly isothermal process, adjustments were needed because it is not possible to have a perfectly isothermal process in practice. To ensure the compression process in practice is as close as possible to an isothermal process, all the work done by mechanical forces was transferred to a heat reservoir (thermal bath). This heat will be injected back into the compressed air flows before it expands within the turbine to release the stored energy.

6.2.6.2 Storage-Integrated Vortex Hydrokinetic Energy Converter (SAVER)

Storage integrAted Vortex hydrokinetic Energy ConverteR (SAVER), a modified version of the device described by Vasel-Be-Hagh et al. [22], uses the best features of two existing technologies: (i) vortex hydrokinetic energy harvester [23,24] and (ii) underwater compressed air energy storage [21]. The energy harvesting module of the proposed device is a device that converts kinetic energy of vortex-induced vibrations (VIV) into electrical power. VIV is a naturally occurring phenomena that happens to solid bodies in cross-flow. As water flows over a bluff body, vortices shed off of the surface in what is known simply as *vortex shedding.* Vortex shedding causes the bluff body to oscillate.

6.2.6.2.1 How SAVER Works

The process flow diagram of SAVER is shown in Figure 6.4. The system consists of five main subsystems including the air compression unit, transmission pipeline unit, expansion unit, heat recovery unit, and most importantly, the submerged

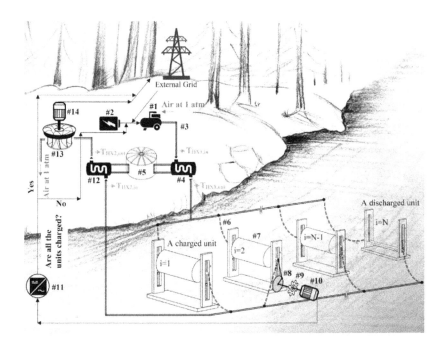

FIGURE 6.4 Overview of the proposed power-generating storage plant: (#1) compressor unit, (#2) starter supply, (#3) rigid compressed air pipes, (#4) heat exchanger no. 1, (#5) heat tank, (#6) flexible compressed air lines, (#7) a charged accumulator-converter (A-C), (#8) crankshaft, (#9) gearbox, (#10) generator driven by oscillating A-Cs, (#11) AC-DC converter, regulator, inverter, and filter, (#12) heat exchanger no. 2, (#13) turboexpander driven by compressed air released to quickly respond to high demands, and (#14) generator driven by turboexpander. To evade the effects of the upstream units on the vortex-induced vibrations (VIV) of the downstream ones, each A-C should be placed in a staggered pattern off the wake of the upstream.

accumulator-converter (A-C) unit. In the proposed system, energy is stored in the form of potential energy of compressed air underwater, where a charged compressed air accumulator also plays a second role as an active energy converter. Highly durable flexible cylindrical lift bags are suitable choices to perform as the accumulator-converters (A-Cs) for this hybrid system. See Figure 6.4 for a plant of N SAVER units. Four sample units, that is, 1st, 2nd, (N-1)th and Nth, are shown in the figure. A single SAVER unit with more details is displayed in Figure 6.5. The surplus electrical energy generated in the off-peak hours in addition to the power harvested by charged A-Cs is used to run the compression unit (Figure 6.4: #1). If there is no surplus power from the external grid, and all A-Cs are discharged, the system can be initiated using a starter battery (Figure 6.4: #2). This compresses atmospheric air and stores it in sealed underwater flexible accumulator-converters (A-Cs) at much higher pressures. This is a major modification in comparison with conventional underwater CAES where the accumulator balloons are all open from the bottom and the compressed air would be discharged into the water if the air pressure exceeds hydrostatic pressure at the installation depth. This broadens the application of underwater CAES beyond very deep waters. The terminal pressure depends on the strength of material used for manufacturing the A-C balloons. The temperature of air increases during the compression process, hence, using a plate heat exchanger (Figure 6.4: #4); the heat added to the air will be taken and stored in an insulated heat tank (Figure 6.4: #5) before transferring the compressed air into underwater A-Cs where compressed air will reach a thermal equilibrium with cold water. Storing heat of compressed air in a heat tank pushes this process toward an adiabatic condition and significantly reduces the thermal losses, which

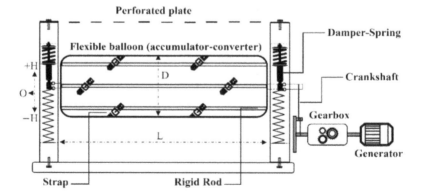

FIGURE 6.5 A charged balloon performing as the accumulator-converter of the proposed Storage-integrAted Vortex hydrokinetic Energy converteR (SAVER). Rigid rods are strapped to the flexible pressurized balloon to prevent deflection under hydrostatic pressure and dynamic loads. The perforated plate installed at the top would ensure that in case of bursting of the balloon, which is very unlikely, any generated vortex ring is dispersed into smaller spherical bubbles before it gains any momentum. (See Bernitsas, M., et al., *J. Offshore Mech. Arctic Eng.*, 131, 1–13, 2009 for further details.)

leads to higher efficiencies. Cold compressed air is stored in the A-Cs (Figure 6.4: #7) through a rigid (Figure 6.4: #3) and flexible (Figure 6.4: #6) piping until all A-Cs are charged. A-Cs are always locked during the charging process to avoid technical complications; hence, they do not contribute to the plant power generation until they are fully charged with compressed air at a predefined terminal pressure. Once an A-C is fully charged, it is unlocked to start its linear oscillations in response to the VIV from passing water currents. This linear motion is converted to rotary motion using a crankshaft mechanism (Figure 6.4: #8), the speed of which is regulated using a gearbox (Figure 6.4: #9). The crankshaft allows the use of electric machines that do not use rare Earth magnets and can be modularly installed with relative ease. It also offers the capability of placing the generators above the water for easier maintenance. The rotary motion drives a generator (Figure 6.4: #10). The generator output is passed through an AC-DC converter consisting of an electronic rectifier, regulator, inverter, and filter (Figure 6.4: #11) to produce a 60 Hz power signal that could interface with the grid. Electrical power produced by this generator is transmitted either to the compressor unit (if there are still uncharged A-Cs) or to the external grid (if all available A-Cs are charged). If more power is urgently required during peak hours, A-Cs can be locked one by one, and the stored compressed air can be quickly discharged through a turbine (Figure 6.4: #13) with the help of hydrostatic pressure of water that continuously acts on compressed air stored in flexible A-Cs. The electrical power generated using released compressed air is then transmitted to the grid to meet the demand. Note that the delivered cold compressed air from A-Cs is first heated by passing through the plate heat exchangers (Figure 6.4: #12) before entering the turbine. Heat exchanger liquid supplied from the heat tank (Figure 6.4: #5) flows in the reverse direction to heat this cold air (60% ethylene glycol solution can be used as the heat exchanger liquid).

6.2.6.2.2 Advantages of SAVER

The proposed device mixes two existing cost-effective, simple, environmentally friendly, and scalable technologies (underwater compressed air energy storage [UWCAES] and vortex hydro energy harvester [VHEH]) to develop a highly efficient device, that:

1. Costs in the same order of traditional hydro energy harvester plants (in utility scale), while, unlike traditional harvesters, SAVER is both a power generator and an energy storage device
2. Does not suffer from intermittency and unpredictability due to its integrated storage module
3. Operates at currents as slow as 1 m/s
4. Operates at waters as shallow as 5 m, making it a more feasible option than conventional UWCAES (this will expand the renewable resource capacity of the market, as there are more potential areas that can host this device)
5. Requires much shorter piping and wiring as it can be installed close to the shoreline, making the installation and maintenance operations much easier and cheaper.

6.2.6.2.3 Efficiency Calculations

Energy calculations for a typical SAVER unit are presented. The numerical calculations presented in this section are based off deploying the unit at the Bourne Tidal Test Site, located in the Cape Cod Canal. Characteristic values for this typical unit are given in Table 6.6. Ultimately, the roundtrip efficiency for a single unit is obtained. While in practice, the SAVER unit would be used in an array system and one needs to take into account the negative effects of hydro-mechanic wakes on the efficiency of the downstream units. The following calculations should give an idea of what should be reasonably expected.

Step 1: Calculating the storage capacity of a single SAVER unit E_{Stored} **[MJ]**

The storage capacity of a single SAVER unit is calculated via the following equation:

$$E_{Stored} = p_2 V \, ln \frac{p_1}{p_2}. \tag{6.3}$$

By substituting the data provided in Table 6.6 into Equation (6.3), the amount of energy that can be stored in the storage module of each unit is calculated as $E_{Stored} = 7.72$ MJ.

Step 2: Calculating the amount of energy that can be delivered from storage module [MJ]

The amount of energy that can be delivered from the storage module is calculated as $E_{out} = \eta_{round-trip} E_{Stored}$. Roundtrip efficiency of a conventional UWCAES unit is reported to be approximately 65% [26]. Although

---

TABLE 6.6

Characteristic Values for a Typical Unit

Initial pressure at the inlet, of the compressor	$p_1 = 1$ atm $= 101,325$ Pa
Atmospheric air temperature	$T_1 = 20°C = 293$ K
Final pressure at the accumulator-converters (A-Cs)	$p_2 = 6$ atm $= 6 \times 101,325$ Pa
Water temperature	$T_2 = 10°C = 283$ K
Compressor unit volumetric flow rate (CFM)	$Q = 35$ CFM $= 0.0165$ m³/s
Volume of each accumulator-converter	$V = 7.1$ m³
Length of each cylindrical accumulator-converter balloon	$L = 4$ m
Diameter of each cylindrical accumulator-converter balloon	$L = 1.5$ m
Amplitude of the oscillations	$H = 1$ m
Current speed	$U = 1$ m/s

this value changes with isentropic efficiencies of used turbines and compressors. Using heat exchangers that capture more thermal energy would also increase system performance, although to a lesser extent compared to the impact of isentropic efficiencies of turbine and compressor units. The authors, however, calculated the target level of performance for a single SAVER unit assuming that the roundtrip efficiency of the storage module of this device is $\eta_{round-trip}$ = 55%, approximately 10% less than what is reported in the literature to account for uncertainties and other potential inefficiencies that might cause a lower roundtrip efficiency. Therefore, E_{out} = 3.68 MJ.

Step 3: Calculating the charge/discharge period(s)

Charge: Mass of compressed air stored in a charged SAVER unit is calculated as $m = p_2 V / RT_2$. Intake mass flow rate at the compressor unit is calculated as $\dot{m} = \rho Q = p_1 / RT_1 Q$. Therefore, charge period can be calculated as:

$$\Delta_{charge} = \frac{m}{\dot{m}} = \frac{p_2 V / RT_2}{p_1 Q / RT_1} = \left(\frac{p_2}{p_1} \right) \left(\frac{T_1}{T_2} \right) \left(\frac{V}{Q} \right). \tag{6.4}$$

In other words, charge period depends on the pressure ratio (pressure of stored air divided by atmospheric pressure), temperature ratio (atmospheric air temperature divided by water temperature, assuming thermal equilibrium between water and stored air), storage volume of each SAVER unit, and compressor flow rate in CFM. Therefore, using data given in Table 6.6:

$$\Delta_{charge} = 2673 \text{ s} = 45 \text{ min}$$

Discharge: Discharge period depends on the discharge rate, which is controlled by the operator who sets it based on the instantaneous demand. The discharge period is usually longer than the charge period.

Step 4: Calculating energy that can be harvested by a single charged SAVER unit

Total hydrokinetic power available in water current is calculated as:

$$P = \frac{1}{2} \rho A U^3 \tag{6.5}$$

where ρ, A, and U are density of water, swept area of the oscillating A-Cs, and current speed, respectively. Swept area is calculated as $A = WL$, where

L represents length of each A-C unit and W is defined as $W = H + D$, in which H is the amplitude of the oscillations and D is the diameter of each cylindrical balloon.

According to [27] energy conversion efficiency of conventional vortex hydro energy harvester is approximately $\eta_{conv} = 22\%$. However, the authors assumed an energy conversion efficiency of $\eta_{conv} = 17\%$, which is 5% less than values reported in the literature to account for any potential additional losses. Therefore, power harvested by the single SAVER unit described in Table 6.6 at a current speed of 1 m/s is:

$$P_{harvested} = \eta_{conv} \frac{1}{2} \rho A U^3. \qquad (6.6)$$

$$P_{harvested} = 850 \ \text{W}$$

Due to complications caused by the variation of mass during charge/discharge periods, which causes a continuous change in the optimal required set of spring/damper system for the device, the cylinder is locked during the charging/discharging process. Power production starts when the cylinder is fully charged and continues until there is a need for quick power release by quickly discharging compressed air stored in the cylinder (beginning of discharge process).

Step 5: Calculating roundtrip efficiency

To calculate the roundtrip efficiency of the integrated system, it is critical to know how much energy (J) was produced by power harvester module of the SAVER. The amount of energy produced by the power harvester module of SAVER directly depends on the period between the end of charge process and the beginning of discharge process. The longer this period of power production lasts, the more energy can be harvested for making up for losses. Here the roundtrip efficiency of a single SAVER unit is calculated for two different cases: (i) assume the power production period is in the same order of the charge period, and (ii) the power production period is one order of magnitude larger than charging period.

Sample 1: Assume the power production period (i.e., the period between the end of charging and the beginning of discharging) is in the same order of the charge period. Then, using the charge period calculated at step 3, and power production rate calculated at step 4, the harvested energy is:

$$E_{harvested} = P_{harvested} \, \Delta_{production} = 850 \times 2673 = 2.27 \ \text{MJ}$$

Therefore, the roundtrip efficiency of the unit is:

$$\eta_{round\text{-}trip} = \frac{E_{out} + E_{harvested}}{E_{stored}} 100\% = 83.5\% \tag{6.7}$$

Sample 2: Normally, the power production period is one order of magnitude larger than charging period, approximately equal to the off-peak duration, which varies between 7 and 10 hours depending on the season. Assuming power production period equal to approximately 8 hours:

$$E_{harvested} = P_{harvested}\,\Delta_{production} = 850 \times (8 \times 3600) = 24.48 \text{ MJ}$$

Leading to:

$$\eta_{round\text{-}trip} = \frac{E_{out} + E_{harvested}}{E_{stored}} 100\% = 367\% \tag{6.8}$$

Therefore, not only are all losses compensated, but also an additional amount of electrical energy equal to $2.67\ E_{stored}$ is produced during an 8-hour period of operation of the proposed SAVER.

6.3 GENERAL COMPARISON

Although it is difficult to have an all-inclusive comparison of every technology presented within this chapter, it is important to at least have some comparison with similar characteristics that were provided. All technologies discussed in this chapter have been put up together in Table 6.7. Many of the discussed technologies did not provide an efficiency rating because many were new ideas just being presented to the engineering community. While the efficiency (if provided) was added to Table 6.7, to get a more inclusive comparison, the main advantages and negative features were added, along with a column for location restrictions.

TABLE 6.7
General Characteristic Comparison

Technology	Benefits	Disadvantages	Efficiency (%)	Location Restrictions
Hydro Pumped Storage (HPS)				
HPS + Wind Turbine via open wells [10]	+ Small scale + Quick response + Very uniform output + Powers a house for multiple hours without any wind	– Immature technology – Losing energy in the expense of addressing intermittence	Not provided	Requires open wells
HPS + Wind-Diesel via flood water [12]	+ Increases penetration of wind to 100% + Alleviates other issues like flooding	– Very expensive – Location dependent	Not provided	Large body of water
Flywheel				
Flywheel + Batteries + Wind/Solar [13]	+ Reduced number of batteries needed + Long life	– Needs battery storage – Potential danger with failure	75	Ample area
Batteries				
Battery + Wind [15]	+ Large daily saving ($) + Less reliant on diesel generators	– Large scale setup – Bad for the environment – Batteries are expensive	Not provided	Not applicable
Buoyancy-Based Energy Storage (BBES)				
BBES + Wind [17]	+ High scalability + Very high storage capabilities	– High capital cost – Difficult to implement	83	Deep water
Compressed Air Energy Storage (CAES)				
CAES + Floating Solar [19]	+ Lowered environmental impact + Lowered cost and longer life	– Difficult to implement	80	Extensive water area
CAES + VIV-Based Harvesters	+ Environmentally friendly + Scalable + Cost effective + Efficient	– Difficult to implement	Depends on the storage period	Water

REFERENCES

1. B. Sørensen, "A history of renewable energy technology," *Energy Policy*, vol. 19, no. 1, pp. 8–12, 1991.
2. P. Driessen, "Greens shackle national security – And renewable energy," *Energy and Environment*, vol. 22, no. 4, pp. 425–428, 2011.
3. T. Hamed and L. Bessler, "Energy security in Israel and Jordan: The role of renewable energy sources," *Renewable Energy*, vol. 135, pp. 378–389, 2019.
4. "energy.gov," US Department of Energy, January 2017. Available: https://www.energy.gov/policy/articles/valuation-energy-security-united-states. (Accessed May 2019).
5. J. Krane, "Climate change and fossil fuel: An examination of risks for the energy industry and producer states," *MRS Energy & Sustainability*, vol. 4, p. E2, 2017.
6. J. Lelieveld, K. Klingmüller, A. Pozzer, R. Burnett, A. Haines and V. Ramanathan, "Effects of fossil fuel and total anthropogenic emission removal on public health and climate," *Proceedings of the National Academy of Sciences of the United States of America*, vol. 116, no. 15, pp. 7192–7197, 2019.
7. M. Kirli and M. Fahrioğlu, "Sustainable development of Turkey: Deployment of geothermal resources for carbon capture, utilization, and storage," *Energy Sources, Part A: Recovery, Utilization and Environmental Effects*, vol. 41, no. 14, pp. 1739–1751, 2019.
8. S. Shafiee and E. Topal, "When will fossil fuel reserves be diminished?" *Energy Policy*, vol. 37, no. 1, pp. 181–189, 2009.
9. V. Shahir, C. Jawahar, P. Suresh and V. Vinod, "Experimental investigation on performance and emission characteristics of a common rail direct injection engine using animal fat biodiesel blends," *Energy Procedia*, vol. 117, pp. 283–290, 2017.
10. H. Daneshi, A. Srivastava and A. Daneshi, "Generation scheduling with integration of wind power and compressed air energy storage," in: *IEEE PES T D 2010*, New Orleans, LA, April 2010, pp. 1–6.
11. B. Pali and S. Vadhera, "A novel pumped hydro-energy storage scheme with wind energy for power generation at constant voltage in rural areas," *Renewable Energy*, vol. 127, pp. 802–810, 2018.
12. A. Safi and F. Basrawi, "Evaluation of the wind pumped hydropower storage integrated flood mitigation system," *IOP Conference Series: Materials Science and Engineering*, vol. 342, p. 012101, 2018.
13. G. N. Prodromidis and F. A. Coutelieris, "Simulations of economical and technical feasibility of battery and flywheel hybrid energy storage systems in autonomous projects," *Renewable Energy*, vol. 39, no. 1, pp. 149–153, 2012.
14. C. Woodford, "How do supercapacitors work?" August 2018. Available: https://www.explainthatstuff.com/how-supercapacitors-work.html. (Accessed 2018).
15. S. Tiwari, B. Dwivedi and M. Dave, "Optimized generation scheduling of thermal generators integrated to wind energy system with storage," *International Journal of Renewable Energy Research*, vol. 8, no. 2, pp. 692–701, 2018.
16. M. Fazeli, "Wind generator-energy storage control schemes for autonomous grid," July 2011. Available: http://eprints.nottingham.ac.uk/11698/. (Accessed 2018).
17. K. Bassett, R. Carriveau and D.-K. Ting, "Integration of buoyancy-based energy storage with utility scale wind energy generation," *Journal of Energy Storage*, vol. 14, pp. 256–263, 2017.
18. B. Cheung, N. Cao, R. Carriveau and D.-K. Ting, "Distensible air accumulators as a means of adiabatic underwater compressed air energy storage," *International Journal of Environmental Studies*, vol. 69, no. 4, pp. 566–577, 2012.

19. R. Cazzaniga, M. Cicu, M. Rosa-Clot, P. Rosa-Clot, G. Tina and C. Ventura, "Compressed air energy storage integrated with floating photovoltaic plant," *Journal of Energy Storage*, vol. 13, pp. 48–57, 2017.

20. J. Wang, K. Lu, L. Ma, J. Wang, M. Dooner, S. Miao, J. Li and D. Wang, "Overview of compressed air energy storage and technology development," *Energies*, vol. 10, no. 7, p. 991, 2017. doi:10.3390/en10070991.

21. B. Cheung, R. Carriveau and D. Ting, "Multi-objective optimization of an underwater compressed air energy storage system using genetic algorithm," *Energy*, vol. 74, pp. 396–404, 2014.

22. A. Vasel-Be-Hagh, R. Carriveau and D. S.-K. Ting, "Underwater compressed air energy storage improved through Vortex Hydro Energy," *Sustainable Energy Technologies and Assessments*, vol. 7, pp. 1–5, 2014.

23. M. Bernitsas, K. Raghavan, Y. Ben-Simon and E. Garcia, "VIVACE (Vortex Induced Vibration Aquatic Clean Energy): A new concept in generation of clean and renewable energy from fluid flow," *Journal of Offshore Mechanics and Arctic Engineering*, vol. 130, no. 4, p. 041101, 2008.

24. K. Raghavan and M. Bernitsas, "Experimental investigation of Reynolds number effect on vortex induced vibration of rigid circular cylinder on elastic supports," *Ocean Engineering*, vol. 38, no. 5, pp. 719–731, 2011.

25. A. Vasel-Be-Hagh, R. Carriveau and D.-K. Ting, "A balloon bursting underwater," *Journal of Fluid Mechanics*, vol. 769, pp. 522–540, 2015.

26. B. Cheung, R. Carriveau and D.-K. Ting, "Parameters affecting scalable underwater compressed air energy storage," *Applied Energy*, vol. 134, pp. 239–247, 2014.

27. M. Bernitsas, Y. Ben-Simon, K. Raghavan and E. Garcia, "The VIVACE converter: Model tests at Reynolds numbers around 10^5," *OMAE 2006; and Journal of Offshore Mechanics and Arctic Engineering*, vol. 131, pp. 1–13, 2009.

7 Emerging Water-Energy-Food Nexus Lessons, Experiences, and Opportunities in Southern Africa

Tafadzwanashe Mabhaudhi, Sylvester Mpandeli, Luxon Nhamo, Vimbayi G. P. Chimonyo, Aidan Senzanje, Dhesigen Naidoo, Stanley Liphadzi, and Albert T. Modi

CONTENTS

7.1   INTRODUCTION

The water-energy-food (WEF) nexus is a systems approach that improves resource management, use, and distribution, managing cross-sectoral synergies and trade-offs in a holistic way (Albrecht et al., 2018). This systems approach has become part of the modern development norm as sectoral approaches are now viewed as unsustainable (Mpandeli et al., 2018; Terrapon-Pfaff et al., 2018). The popularity of the approach is motivated by the increasing demand of depleting resources perpetuated by increasing demand from a growing population, uneven distribution of resources, and high climatic variability and change (Nhamo et al., 2019b). Consequently, the WEF nexus now forms part of a global research agenda aimed at sustainable development, systems thinking, and transdisciplinary research (Nhamo et al., 2019b). Formal published evidence of the three-way mutual interactions among the WEF nexus components only started emerging in 2008 (Hellegers et al., 2008). The approach then shot into prominence at the World Economic Forum held in Bonn in 2011 where it was promoted as a tool to achieve and manage sustainable economic development and has since grown into an important global research agenda (Scott et al., 2015).

As a polycentric approach, the WEF nexus is used either as an analytical tool, a conceptual framework, and a discourse or as part of a decision support system (DSS). As an analytical tool, the nexus systematically uses quantitative and qualitative methods to understand the interactions among WEF resources (Nhamo et al., 2019b). As a conceptual framework it leverages an understanding of WEF linkages to promote coherence in policy making and enhances sustainability to promote cross-sectoral approaches (Nhamo et al., 2019b). As a discourse it is a tool for problem framing and promoting cross-sectoral collaboration, and as a DSS it is used to inform resource planning and management decisions (Nhamo et al., 2019b). These WEF nexus niches augur well with the transboundary nature and uneven distribution of resources in southern Africa. For southern Africa, the approach has potential to enhance climate change adaptation and adaptive capacity, promote regional integration, build resilience, and improve the livelihoods of people (Mpandeli et al., 2018; Nhamo et al., 2018). The WEF nexus approach is a pathway for understanding complex and dynamic interlinkages between issues related to water, energy and food security, it is strongly linked to Sustainable Development Goals (SDGs), particularly SDGs 2 (zero hunger), 6 (clean water and sanitation), and 7 (affordable and clean energy) (Mabhaudhi et al., 2018a).

Despite the reported value of the WEF nexus approach, it has remained as a rhetoric ambition without proper implementation, monitoring, and evaluation guidelines to direct policy and decision making (Leck et al., 2015; Rees, 2013). A substantial amount of literature has been published highlighting the importance of the WEF nexus as a conceptual framework and as a discourse but evidently lacking robust analytical tools (Liu et al., 2017; Nhamo et al., 2018; Terrapon-Pfaff et al., 2018). Critics of the WEF nexus base their arguments on the failure of the approach to offer real-world solutions and become a fully fledged operational framework (Terrapon-Pfaff et al., 2018). The approach has lacked tools to evaluate synergies and trade-offs in an integrated way and provide decision support tools to inform policy making and implementation across the WEF sectors (Howells et al., 2013; Liu et al., 2017). It is only recently when analytical tools have been developed to look into the three resources in a holistic way (Nhamo et al., 2019b). Previous tools lacked the main attributes of a nexus analytical framework because most remained either theoretical or still maintained a sectoral approach to resource development, use, and management (Albrecht et al., 2018; McGrane et al., 2018).

As research interest around the WEF nexus has been evolved, regional and national WEF nexus consortiums have been formed within the framework of multi-stakeholder dialogues to look into developing analytical tools (Daher et al., 2018). The most outstanding of these consortiums are (i) the WEF Nexus Group formed by international organizations and universities at the World Water Week in Stockholm in 2016 (http://wefnexusgroup.org/wwf8); (ii) the Nexus Network a grouping of mainly international universities was formed in 2014 to support WEF nexus transdisciplinary research and to create meaningful links between communities of researchers, policy makers, business leaders, and practitioners (https://thenexusnetwork. org); and (iii) the Nexus Platform, formed in 2011 as an independent information and facilitating platform funded by the Federal Ministry of Economic Cooperation and Development and the European Union (www.water-energy-food.org/nexus-platform-the-water-energy-food-nexus). These high-level global platforms signify how the WEF nexus agenda has attracted global attention as a pathway for sustainable development.

Within southern Africa, the approach has also gathered a lot of interest as evidenced by the increasing availability of literature on WEF nexus dedicated to the region (Mabhaudhi et al., 2016; Mabhaudhi et al., 2018b; Mpandeli et al., 2018; Nhamo et al., 2018). At the Southern Africa Development Community (SADC), the SADC Secretariat has been spearheading WEF nexus research through the Global Water Partnership (GWP) since the *6th Multi-stakeholder Water Dialogue* held in Lusaka, Zambia in 2013, which focused on the WEF sector linkages (Mabhaudhi et al., 2018a). There is a lot of WEF nexus research going on in the region being spearheaded by watercourse commissions such as Limpopo River Basin Watercourse Commission (LIMCOM) and the Zambezi River Basin Watercourse Commission (ZAMCOM), as well as universities and research organizations. The SADC Secretariat has produced the WEF Nexus Action Plan through the Regional Strategic Action Plan (RSAP) on Integrated Water Resources Development and Management (SADC, 2016) that recognizes the role of the nexus in adapting to the challenges posed by population growth and climate variability and change, as well as in optimizing resource use. The WEF Nexus Action Plan is a regionwide WEF nexus operational framework aimed to support the attainment of regional goals and targets that include regional integration, poverty alleviation, and improve the livelihoods of people. The adoption of the WEF nexus approach in southern Africa is precipitated by the need to ensure simultaneous securities of WEF resources, a topical issue that has dominated the development agenda of southern African, centered on improving livelihoods, building resilience, and regional integration (Cervigni et al., 2015; Davidson et al., 2003).

This chapter provides an overview of progress on the WEF nexus, giving a focus on southern Africa, assessing regional water, energy and food resources, exploring opportunities for the nexus in promoting cross-sectoral policy harmonization and development of sustainable climate change adaptation strategies, as well as linking the WEF nexus to related SDGs. The chapter discusses regional resource endowment, policy and institutional arrangements, and proposes a governance framework for WEF nexus adoption. The challenge of resource scarcity is of particular concern in southern Africa due to dependence on climate-sensitive sectors of agriculture and energy, which heavily depend on water resources. For the region, the WEF nexus approach has potential to integrate strategies aimed at adapting to the challenges brought about by population growth, increased urbanization, increased consumption demands because of improved standards of living, and climate variability and change.

7.2   THE WEF NEXUS IN THE CONTEXT OF THE SOUTHERN AFRICA

Sixty percent of the population of the SADC region live in rural areas and depend on natural systems for their livelihoods as well as relying on rainfed agriculture (Mabhaudhi et al., 2018a; Nhamo et al., 2019c). Most of the rural inhabitants lack access to clean energy, water, and sanitation and are faced with food insecurity (Mabhaudhi et al., 2016; Nhamo, 2015). The underdevelopment and increasing vulnerabilities of the region are caused, in part, by sectoral approaches in resource development, use, and management (Nhamo et al., 2018) The "silo" approach in resource management practiced at national level inadvertently contributes to the region's failure to meet its development targets. The essence of the WEF nexus is its capability to address the challenge of sectoral management of resources through the harmonization of institutions and policies, as well as setting targets and indicators to implement and assess resource management for sustainable development. For the SADC region, the WEF nexus presents opportunities for integrated resource management for regional integration and inclusive socioeconomic development and security, mainly because resources are generally transboundary in nature (Mabhaudhi et al., 2016). In the SADC, the WEF nexus could be valuable when it comes to promoting inclusive development and transforming vulnerabilities into resilient communities.

The resource rich fifteen transboundary river basins (IRBs) in the SADC region (Figure 7.1) present opportunities for coordinated and sustained growth and ensure socioeconomic security through the WEF nexus. Riparian countries could achieve short- and long-term benefits through an integrated and coordinated operation of existing and planned hydropower facilities, cooperative flood management, and irrigation development (Mabhaudhi et al., 2016). According to the World Bank, cooperation among riparian countries of the Zambezi Basin has the potential to bring reasonable balance between hydropower and irrigation investment that could result in stable energy generation of some 30,000 Gigawatt hours (GWh)/year and unlock 774,000 ha of irrigated land (The World Bank, 2010). Currently, most large dams in the region are underused because they were originally designed for single purposes. However, some like the Itezhi-Tezhi Dam in Zambia, are being redesigned for multipurpose to be used for both hydropower and irrigation (Deines et al., 2013). The Kariba Dam was also originally commissioned only for hydropower generation but is now used for aquaculture, urban water supply, tourism, support to national parks and wildlife, lake transportation, and mining activities (Nyikahadzoi et al., 2017).

Evidently, such scenarios create new economic opportunities that promote inclusive growth, job creation, and sustainable development. However, these are only a few cases that highlight WEF nexus synergies. Adopting and implementing the WEF nexus at the regional level in the SADC will minimize conflict and unnecessary tension among member states and allow inclusive development and investment, as the nexus promote cross-sectoral policy linkages. As the WEF nexus components are sensitive to climate variability and change, its adoption also promotes resilience-building initiatives and would contribute to the region's climate change response strategies. Thus, the nexus unlocks opportunities for collaboration, boosting regional cooperation, and inclusive development through a set of common targets.

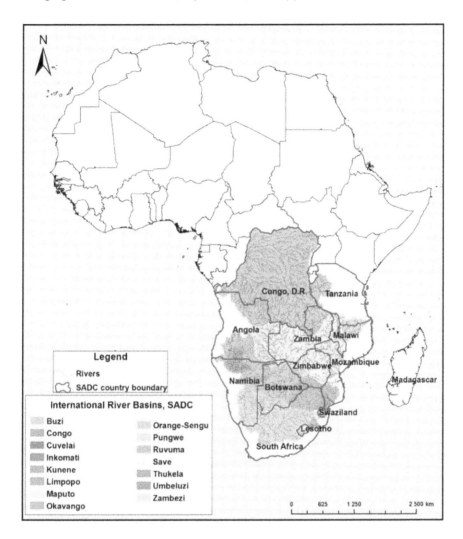

FIGURE 7.1 Southern Africa Development Community (SADC) countries in Africa and the transboundary river basins within the regional economic block.

7.2.1 Resource Endowment in Southern Africa

Rainfall in the SADC region is highly variable, oscillating between 100 and 2,500 mm per annum, and this is indicative of the uneven distribution of hydrological resources across the region. Seventy-five percent of the region is arid or water scarce (physical and economic) (Nhamo et al., 2019a). Total renewable freshwater resources are estimated at 2,300 km^3 per annum of which 7% of water withdrawals is used in agriculture, 16% for domestic and 8% for industrial use (Nhamo, 2015). Although agriculture uses the bulk of freshwater resources, crop production remains very low failing to meet food requirements of a growing population (Mpandeli et al., 2018). Seventy percent of surface water resources are in fifteen transboundary river basins (Figure 7.1). The hydrology of the SADC is, thus characterized by the high number of transboundary river basins (Nhamo et al., 2018), highlighting the importance of watercourses in promoting regional integration and development. Although the southern parts are arid, the Congo, Zambezi and Orange–Senqu basins have the potential to generate significant regional benefits through water transfer and hydropower generation (Stiles and Murove, 2015).

The SADC region is endowed with vast, but underused energy resources, although availability varies from country to country (Stiles and Murove, 2015). The untapped potential of hydropower generation in Angola, the Democratic Republic of Congo, Mozambique, and Zambia has capacity to supply the whole region with electricity (Stiles and Murove, 2015; The World Bank, 2010). However, the main source of energy within the region is biomass as only 24% of the total population and 5% of rural people have access to electricity (Schreiner and Baleta, 2015). Just like water resources, the distribution of energy resources has potential to play an important role in regional integration, inclusive economic development, and poverty eradication (Nhamo et al., 2018). The region currently shares power grids whose electricity is generated from shared watercourses. Regional hydropower potential is estimated at about 1,080 terawatt hours per year (TWh/year), but the current level of exploitation is less than 31 TWh/year (Stiles and Murove, 2015).

At almost 9.9 million km^2, the SADC region's land area is almost a third of the total land mass of the African continent (SADC, 2012). Of this total, 25% is arable, and farming occurs only on 6% of the area (Mabhaudhi et al., 2018b). This in reality means the SADC region has potential for agricultural production and achieving food security at both the national and regional levels. Regrettably, unequal land distribution in the SADC region is a challenge that militates against this potential. This means land reform in the SADC is imperative if agricultural production and rural development are to be realized. However, due to a cautious approach predicated on political, social, and economic considerations, land reform has been woefully slow in the region. Agriculture is the main catalyst for regional economic development as more than 60% of inhabitants depend on it for their livelihoods, providing their subsistence, employment, and income (Mabhaudhi et al., 2018b). The performance of the agriculture sector, therefore significantly impacts on economic growth, poverty reduction, and food security as the sector accounts for close to 17% of the region's gross domestic product (GDP), and the contribution increases to more than 28% when middle income countries are excluded (SADC, 2014a). Despite its importance to regional economic growth, agricultural growth rates remain very low and highly variable averaging only 2.6% per annum. This falls below the regional target of 7% per annum. Agriculture growth rates have almost been at par with population growth rates of 2.4% (Chilonda and Minde, 2007) resulting in food insecurity. Current annual performance of the sector has been insufficient to significantly contribute to regional economic growth and address food and nutrition security issues in the region. Dependence on rainfall for agriculture increases the vulnerability of the region to the vagaries of extreme climatic events, high climate variability, and change. Yet, land with irrigation potential is approximately 20 million ha, of which only 3.9 million ha is equipped for irrigation, accounting for about 6.6% of cultivated area (SADC, 2015).

7.2.2 Regional Institutions and Policies Related to the WEF Nexus

The SADC Secretariat has formulated legal frameworks for the region that can provide the political will to implement the WEF nexus. There are institutions and policies in place to oversee and direct water, energy, and agriculture resources, but they still relate to each sector. There is, therefore, need to harmonize policies in the region and embrace the cross-sectoral approach of the WEF nexus. The SADC treaty is the overarching framework for the region, whose objective is to achieve economic development, peace and security, and growth and also to alleviate poverty and improve the livelihoods of the people, all of which are achieved through regional integration (SADC, 2011). To date, the region has ratified the following nexus-related institutions and policies:

1. The Regional Strategic Action Plan IV (RSAP IV) (SADC, 2015) is based on the SADC Water Policy and Strategy that aims to achieve an equitable and sustainable utilization of water for social and environmental justice, regional integration, and economic benefit for present and future generations, emphasizing infrastructure development and water resource management for food security in the water-food nexus.
2. The SADC protocol on shared watercourses (SADC, 2000) fosters closer cooperation for judicious, sustainable, and coordinated management, protection, and use of shared watercourses, and advancement of SADC's agenda of regional integration and poverty alleviation.
3. The Southern African Power Pool (SAPP) is a grouping that was established in 1995 guided by the Protocol on Energy (SADC, 1996), which highlights the development and updating of a regional electricity master plan, and the development and use of electricity in an environmentally sound manner, while emphasizing the need for universal access to affordable and quality services. The mandate of the SAPP is to enhance regional cooperation in power development and trade and to provide nonbinding regional master plans to guide electricity generation and transmission infrastructure delivery.
4. The SADC Regional Agricultural Policy (RAP) (SADC, 2014b) envisages integrated approaches on water resources management and emphasizes the importance of improving agriculture performance to meet the food and water security, as well as attaining sustainable economic development objectives at a regional level. The RAP oversees the upgrading and expansion of water infrastructure for agriculture, data collection for dams, irrigated areas, and irrigation management.
5. The WEF Nexus Action Plan recognizes the role of the nexus in adapting to the challenges posed by population growth and climate variability and change, as well as in optimizing resource use to achieve regional goals and targets.

7.3 CLIMATE CHANGE IMPACTS IN SOUTHERN AFRICA

The largest proportions of vulnerable populations to the impacts of climate change are invariably found on the African continent where chronic water, food, and energy insecurity and malnourishment remain endemic (Nhamo et al., 2018). Southern

Africa is highly vulnerable to climate variability and change and for that reason it is described as a climate change "hot spot" because of its reliance on climate sensitive sectors of agriculture and fisheries, as well as low adaptive capacity (Nhamo et al., 2019a). Water, energy, and food are expected to be the most affected sectors by climate variability and change in the region (Mpandeli et al., 2018).

Climate projections for southern Africa indicate increased physical or economic water scarcity by as early as 2025 (Niang et al., 2008). The Intergovernmental Panel on Climate Change (IPCC) estimates that between 75–250 million and 350–600 million people in Africa will be at risk of increased water stress by 2020 and 2050, respectively (Parry et al., 2007) (Figure 7.2). Rainfall variability threatens the production of more than 80% of agricultural land on the continent, as it is rainfed (Besada and Werner, 2015). Reduced rainfall, coupled with increased temperatures, will reduce (i) the area suitable for agriculture, (ii) the length of growing period, and (iii) yield potential (Nhamo et al., 2019c; Parry et al., 2007). By 2080, rainfall variability and longer dry spells would result in reduction of crop yields, rise in sea levels and coastal and low-lying areas would be affected by floods (Niang et al., 2014). For the distant future (2080–2099), temperature increases of more than 4°C–6°C are likely over the entire South African interior. Such increases will also be associated with drastic increases in the number of heatwave days and very hot days with potentially devastating impacts on agriculture, water security, biodiversity, and human health (Niang et al., 2014). Significant changes are already being experienced in sectors of agriculture, water, energy, biodiversity, and health (Niang et al., 2014). Climate models projects that Africa will be able to provide only 13% of its food requirements by 2050 if no measures are in place to reduce greenhouse gas (GHG) emissions (Niang et al., 2014). Population is anticipated to have increased to 2 billion during the same period. Extreme weather events as caused by climate

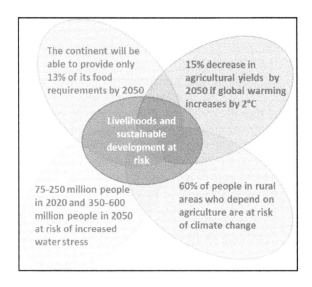

FIGURE 7.2   Highlights of projected climate change risks in sub-Saharan Africa.

change are threatening to derail progress made so far in poverty alleviation, employment, housing, access to and provision of services, food security, and potable water (Thornton et al., 2014). Figure 7.2 highlights some projected climate change risks over southern Africa. Climate change risks threaten human well-being and sustainable development, especially for more vulnerable communities in Africa that rely on natural resources.

Climate change models project that warming over the African continent is occurring at twice the global rate (Engelbrecht et al., 2015; Niang et al., 2014). Without strategies to reduce GHG emissions, temperatures are projected to rise more than 4°C in southern African by 2100. Using the 1981–2000 base period, heat waves have increased by more than 3.5 fold to date in the region (Ceccherini et al., 2017; Dosio, 2017). Such changes in climate regimes, coupled with increased frequency and intensity of floods and droughts, are usually accompanied by health issues. The effects of these extreme weather events transcends mortality and damage to property and crops, but also results in food and water insecurity, spread of disease, and mental health conditions (Machalaba et al., 2015). In most cases, these stressors manifest as drivers of migration, within and between countries (Waldinger, 2015). Extreme weather events are having considerable constraints in southern Africa due to the reliance on natural resources, which are highly sensitive to climate and variability.

Climate change will also result in an increase in vector-borne diseases such as malaria, dengue fever, yellow fever, among others (Githeko et al., 2000; Mpandeli et al., 2018). Health challenges caused by climate change will be highly noticeable in regions experiencing extreme weather events like heatwaves, floods, storms, and fires. Other health risks resulting from climate change will be from changes in regional food yields, disruption of fisheries, loss of livelihoods, and population displacement due to sea-level rise, water shortages, loss of agricultural land, among others (Ali et al., 2017). Water, energy, and food security are closely related to health, as food and water for instance, have a direct impact on human health and the physical conditions of humans has a strong influence on their ability to work (Wlokas, 2008).

7.4 OPPORTUNITIES FOR THE WEF NEXUS IN SOUTHERN AFRICA

The urgency to adopt and implement the WEF nexus is due to an increasing population, changing lifestyle by a growing middle-income group, and industrialization, which are exerting more pressure on WEF resources. Although livelihoods are not explicitly accounted for within the WEF nexus frameworks, a small but growing body of research has highlighted the value of nexus-based approaches for evaluating the effects of development on livelihoods and for promoting sustainable livelihood practices (Biggs et al., 2014, 2015). The high levels of poverty, as well as political instability has triggered economic and political migration within countries and across the region. The rapid urbanization currently being observed across the region has been attributed to young people flocking to urban areas in search of a better life, resulting in a high demand of WEF resources (Gibson and Gurmu, 2012). In addition, the high vulnerability and low productivity evident throughout the region are a tough combination for smallholder farmers who are the majority of the population and are vulnerable to climate change impacts (Nhamo et al., 2019b). Economic and political instability in some areas in the region also fuels migration to more resource-endowed regions, further exerting the demand for WEF resources in receiving areas (Nhamo et al., 2018). These growing pressures on WEF resources require evidence-based solutions of WEF nexus implementation for better resource management to meet the demands of the growing population. We highlight some opportunities for WEF nexus adoption and implementation in southern Africa:

1. *Natural resources management.* Boundaries determined by natural resources go beyond political boundaries and sovereignty. They increase the potential for effective resources management among the WEF sectors (Nhamo et al., 2019a). The transboundary nature of resources, common culture, and similar challenges among countries in southern Africa present an opportunity for regional cooperation through an integrated resource management at regional level. As an example, the southern parts of the region are water scarce and generally do not have enough energy, yet the unexploited resources of the Zambezi and Congo basins are capable of supplying the whole region with enough energy and water (Mabhaudhi et al., 2016). The adoption of the nexus approach has potential to benefit the southernmost countries of the region, which are water scarce.

2. *Irrigation expansion.* Southern Africa has vast but underused agricultural land. Cultivated land accounts for only 6.11% of the total surface area of the region (van Koppen et al., 2015). Land with irrigation potential is about 20 million ha, yet only 3.9 million ha is actually irrigated, which accounts for about 6.6% of the total cultivated area (Nhamo, 2015). The WEF nexus could play an important role in irrigation expansion to provide for food and water security. The Comprehensive Africa Agriculture Development Programme (CAADP) has targeted increasing irrigated area by 100% by 2025 from the baseline value of the year 2000 at an estimated cost of US$37 billion, while infrastructure operation and maintenance required a further US$31 billion (NEPAD, 2014).

3. *Power generation.* Out of a possible 150 GW of hydropower potential, only 12 GW is harnessed (Nhamo et al., 2018). To improve access and meet the 2030 Agenda on supplying clean and affordable energy for all (SDG 7), the region can take advantage of the SADC Power Pool to develop, generate and supply the available and abundant renewable in the region to ensure energy security. The WEF nexus can be used as a decision support tool and resource management as increasing energy supply through hydroelectric, fossil, and biofuels; however, it requires more water supply (Stiles and Murove, 2015).

4. *Simultaneous attainment of water, energy, and food security.* The potential to ensure water, energy, and food security in the region is thus immense, while still operating within the sustainable use of freshwater resources. However, successful implementation of the WEF nexus would require investment and commitment by SADC countries. Other drivers of the WEF nexus such as skilled labor, institutions, and markets are available in the region.

5. *Strengthening of existing institutions.* Institutions and policies that have been established need to be linked to each other, for example, through the Revised Regional Indicative Strategic Development Plan (RISDP 2015–2020), as the region moves toward joint planning to increase synergies between the three nexus sectors. Joint planning would improve hydropower development and irrigation expansion in a more coordinated manner for sustainable development.

6. *Regional integration.* The WEF nexus has been recognized as a key tool for regional integration and development as well as the actual achievement of regional and national SDGs targets (Biggs et al., 2015). However, a lot must be done to unpack the potential of the WEF nexus approach within the SADC region to effectively exploit the many interlinked development opportunities. These would be particularly relevant if the stresses placed on water, energy, and food resources by climate change and other factors are put into consideration. Climate change does not recognize political boundaries nor sovereignty, as was the case with the 2015/2016 drought, which affected the whole region. A regional nexus approach increases regional resilience and also strengthen adaptive capacity against climate change impacts and reduces vulnerability, which permeates the region.

7. *Sustainable development agenda.* The WEF nexus will help framing of future policies since the targets of SDGs 2, 6, 7, 8, and 9 are related to the water, energy, and food sector. Goal 2 accounts for zero hunger, 6 refers to clean water and sanitation, 7 focuses on affordable and clean energy, 8 comprises of affordable work and economic growth, while 9 is aimed at industry, innovation, and infrastructure. Therefore, it is crucial that government planning incorporates a WEF nexus approach for these SDGs as well as illustrate how the SDGs connect with the three primary sectors under consideration. For example, within the context of semi-arid regions, the SDG 2 can only be achieved by improving water productivity and water-use efficiency, hence eradicating food insecurity and improving nutrition. SDG 6 can be achieved by ensuring basic access to water and sanitation and tackling the issue of water scarcity, an undertaking that will require energy. SGD 7 requires the promotion of renewable energy sources, and access to these power sources.

8. *Job creation and economic development.* SDG 8 focuses on job creation, educating the unskilled workforce, as well as working toward a sustainable economic development, while SDG 9 requires improvements in infrastructure, technology, and industrialization.

9. *Climate change adaptation and mitigation.* Adopting and implementing the WEF nexus in southern Africa has potential to ensure water, energy, and food security in times of scarcity as the approach provides policy and decision making with evidence on intervention priorities. The essence of the WEF nexus is that it is a tool for evaluating the performance of resource use and development for the sustainability of livelihoods and an important tool in scenario planning as it is an operational framework for sustainable development (Mpandeli et al., 2018).

7.5 CHALLENGES TO WEF NEXUS ADOPTION

Despite institutional and policy achievements, the region still faces numerous challenges to fully operationalize the WEF nexus as a conceptual framework and a discourse. Over and above the arguments already discussed, there are also a number of factors (technical, political, and social) delaying the adoption and application of the WEF nexus. Some of these factors include-

1. *National versus regional aspirations.* There is little movement on the ground to show WEF nexus implementation as focus is either at the national level or at the basin/regional level. Although projections point to a stronger regional WEF integration, current progress is hindered by policy sections that allow countries to retain the right to develop and implement their own national plans without being obliged to conform to the regional master plan (Nhamo et al., 2018). For instance, some member states are delaying rectifying protocols on shared watercourses because they do not envisage the need at the moment. At the national level, political sovereignty is still strong, which affects genuine cooperation and integration. Regional cooperation in development programs does not remove national sovereignty but fosters integrated economic development and poverty alleviation. Despite this, there is little evidence of commitment by member states to implement the WEF nexus at a regional level. Although national focus may have positives, the shared nature of resources suggests that pulled investments may achieve a greater impact at a regional rather than national scale. Regional countries may also have problems of limited resources to engage and implement the WEF nexus when they might be having more pressing issues such as security consideration, stability, and internal displacement of populations;
2. *Political will.* Despite a belief in the WEF nexus as a resource management tool, there could be low buy-in from member country politicians and technocrats. This is not surprising given that some member countries are yet to sign regional protocols on shared resources (Nhamo et al., 2018). Fatigue by member states to have to kowtow to ever-changing and shifting developmental paradigms; not so long ago it was IWRM, but now it is WEF nexus;
3. *Funding.* Dependency on donor funding to implement the WEF nexus could be another limitation, given that donor funding always comes with operational complexities; and
4. *Availability of expertise and data.* From a technical perspective, it could also be that there is limited technical expertise in the region on the WEF nexus. Lastly, problems associated with data and tools could be a limitation. Most of the tools and models for undertaking WEF nexus analyses were developed in the resources- and data-rich northern countries in Europe and the United States. The same cannot readily be said for the SADC region.

7.6   EMERGING LESSONS AND GAPS FOR SOUTHERN AFRICA—THE WAY FORWARD

The primary goal for the SADC is to foster regional growth and integration which is premised on the realization that the SADC region is unified by a common history, culture, transboundary agreements, and shared natural resources as highlighted in the SADC treaty (SADC, 2011). At the country level, countries have similar goals on poverty alleviation, improving the quality of life for their inhabitants, economic development, and job creation. Regional countries also face similar challenges such as increasing population, increasing rural–urban migration, food insecurity, unemployment, and inequality. Water, energy, and food security are central to the region's plans for sustainable economic development and transformation. It is in this regard that the WEF nexus offers significant opportunities for a coordinated approach to addressing regional challenges and achieving regional goals under climate change.

Research has proposed several climate change adaptation strategies for the region, which include (i) promoting climate smart agriculture, (ii) developing early warning systems (EWS), (iii) IWRM, (iv) promoting renewable energies with low carbon footprint, and (v) increasing monitoring and modeling capacities across each of the WEF sectors (Nzuma et al., 2010). However, these approaches are either water-, energy-, or food-centric and driven by the individual sectors. However, sectoral interventions in resources development risks causing maladaptation through creating imbalances. Based on the advances made thus far with the WEF nexus analytical framework the approach has potential to close the following gaps that slowed economic development and adaptation in the SADC region:

1. The exclusion of policy makers and other stakeholders creates a large gap of knowledge between researchers and policy makers concerning the sustainable use of natural resources.
2. For development of sound WEF nexus policies and acts, there is a need for sustainable resource planning and management beyond government departments through a central authority.
3. Consequently, governance and joint compliance and enforcement structures are needed at local, provincial, and national levels.
4. Interdisciplinary studies between water, energy, and food security are needed to address the knowledge gap, which is required for policy formation.

7.6.1   THE WEF NEXUS FRAMEWORK FOR SOUTHERN AFRICA

Figure 7.3 illustrates an idealistic SADC fit in WEF nexus conceptual framework showing the four fundamental WEF nexus components and their elements (action fields, finance, governance, and innovation). The proposed framework is an initial step toward the adoption and implementation of the WEF nexus at regional level considering similar challenges besetting regional countries as well as the shared resources within the transboundary river basins. The framework touches the issues

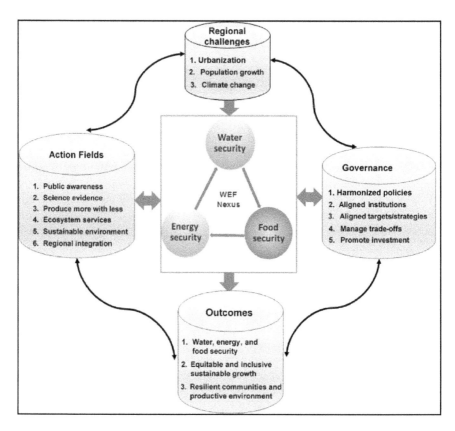

FIGURE 7.3   Regional water-energy-food (WEF) nexus framework for the Southern Africa Development Community (SADC). (From Nhamo, L., et al., *Water*, 10, 18, 2018. With permission.)

common in the region that include challenges and governance and provides the action fields to achieve the desired outcomes through the WEF nexus. The framework emphasizes the role that the WEF nexus could play in regional integration because the region's resources are generally transboundary. The approach could be vital in poverty eradication and resilience building in the advent of climate change and could improve the livelihood of vulnerable people. The WEF nexus framework considers the following elements (Hoff, 2011; Mohtar and Daher, 2016):

1. Strengthening policy and governance to manage the WEF nexus and provide political commitment.
2. Cooperation and commitment by member countries in the implementation of the WEF nexus in shared resources for regional socioeconomic security and poverty eradication.
3. Promotion of public awareness to develop a culture of regional integration and recognition of the role of broader and natural boundaries in regional socioeconomic security and improve the livelihoods of people.

4. Provision of scientific evidence and tools to identify trade-offs between nexus components and support the development of effective, integrative resource allocation strategies.

With the political, will (governance) regional WEF challenges (trends) can be solved by applying nexus assessment models on action plans to achieve desired outcomes (Figure 7.3).

7.7 CONCLUSION

The main challenge currently facing humankind is climate variability and change. In southern Africa, climate variability and change have a multiplier effect on already existing challenges linked to poverty, unemployment, and inequality. This then challenges the region's agenda for inclusive and sustainable development aimed at delivering improved human well-being outcomes for its population. Addressing these challenges requires a systems and holistic approach. The WEF nexus framing draws on a holistic, systems perspectives that recognizes the value of coordinated approaches in resources development, use, and management. The WEF nexus offers opportunities for (i) simultaneous attainment of water, energy, and food securities, (ii) job creation and economic development, (iii) natural resource management, (iv) regional integration and climate change adaptation. However, despite these opportunities, significant challenges to its full-scale adoption and implementation exist. A way forward is to promote and initiate research that generates new data, develops useful metrics and models, and provides useful case studies for informing policy and building political will.

ACKNOWLEDGMENTS

The authors would like to thank the Water Research Commission (WRC), the University of KwaZulu Natal (UKZN), and the International Water Management Institute (IWMI) for supporting the WEF nexus initiative in southern Africa and South Africa and for the support in writing this book chapter.

REFERENCES

Albrecht, T. R., Crootof, A., and Scott, C. A. (2018). The water-energy-food nexus: A systematic review of methods for nexus assessment. *Environmental Research Letters* **13**, 043002.

Ali, S., Liu, Y., Ishaq, M., Shah, T., Ilyas, A., and Din, I. U. (2017). Climate change and its impact on the yield of major food crops: Evidence from Pakistan. *Foods* **6**, 39.

Besada, H., and Werner, K. (2015). An assessment of the effects of Africa's water crisis on food security and management. *International Journal of Water Resources Development* **31**, 120–133.

Biggs, E. M., Boruff, B., Bruce, E., Duncan, J., Haworth, B., Duce, S., Horsley, J., Curnow, J., Neef, A., and McNeill, K. (2014). Environmental Livelihood Security in Southeast Asia and Oceania: A Water-Energy-Food-Livelihoods Nexus Approach for Spatially Assessing Change. White paper, International Water Management Institute (IWMI), Colombo, Sri Lanka.

Biggs, E. M., Bruce, E., Boruff, B., Duncan, J. M., Horsley, J., Pauli, N., McNeill, K., Neef, A., Van Ogtrop, F., and Curnow, J. (2015). Sustainable development and the water–energy–food nexus: A perspective on livelihoods. *Environmental Science & Policy* **54**, 389–397.

Ceccherini, G., Russo, S., Ameztoy, I., Marchese, A. F., and Carmona-Moreno, C. (2017). Heat waves in Africa 1981–2015, observations and reanalysis. *Natural Hazards and Earth System Sciences* **17**, 115–125.

Cervigni, R., Liden, R., Neumann, J. E., and Strzepek, K. M. (2015). *Enhancing the Climate Resilience of Africa's Infrastructure: The Power and Water Sectors.* World Bank Publications, Washington, DC.

Chilonda, P., and Minde, I. (2007). Agricultural growth trends in Southern Africa. In *Policy Brief*, Vol. 1, p. 4. International Water Management Institute (IWMI), Pretoria, South Africa.

Daher, B., Mohtar, R. H., Davidson, S., Cross, K., Karlberg, L., Darmendrail, D., Ganter, C. J., et al. (2018). *Multi-stakeholder Dialogue: Water-Energy-Food (WEF) Nexus and Implementing the SDGs.* International Water Resources Association (IWRA), Paris, France.

Davidson, O., Halsnæs, K., Huq, S., Kok, M., Metz, B., Sokona, Y., and Verhagen, J. (2003). The development and climate nexus: The case of sub-Saharan Africa. *Climate Policy* **3**, S97–S113.

Deines, A. M., Adam Bee, C., Katongo, C., Jensen, R., and Lodge, D. M. (2013). The potential trade-off between artisanal fisheries production and hydroelectricity generation on the Kafue River, Zambia. *Freshwater Biology* **58**, 640–654.

Dosio, A. (2017). Projection of temperature and heat waves for Africa with an ensemble of CORDEX regional climate models. *Climate Dynamics* **49**, 493–519.

Engelbrecht, F., Adegoke, J., Bopape, M.-J., Naidoo, M., Garland, R., Thatcher, M., McGregor, J., Katzfey, J., Werner, M., and Ichoku, C. (2015). Projections of rapidly rising surface temperatures over Africa under low mitigation. *Environmental Research Letters* **10**, 085004.

Gibson, M. A., and Gurmu, E. (2012). Rural to urban migration is an unforeseen impact of development intervention in Ethiopia. *PLoS One* **7**, e48708.

Githeko, A. K., Lindsay, S. W., Confalonieri, U. E., and Patz, J. A. (2000). Climate change and vector-borne diseases: A regional analysis. *Bulletin of the World Health Organization* **78**, 1136–1147.

Hellegers, P., Zilberman, D., Steduto, P., and McCornick, P. (2008). Interactions between water, energy, food and environment: Evolving perspectives and policy issues. *Water Policy* **10**, 1–10.

Hoff, H. (2011). Understanding the Nexus. Background paper for the Bonn 2011 Conference: the Water, Energy and Food Security Nexus. Stockholm Environment Institute, Stockholm.

Howells, M., Hermann, S., Welsch, M., Bazilian, M., Segerström, R., Alfstad, T., Gielen, D., Rogner, H., Fischer, G., and Van Velthuizen, H. (2013). Integrated analysis of climate change, land-use, energy and water strategies. *Nature Climate Change* **3**, 621.

Leck, H., Conway, D., Bradshaw, M., and Rees, J. (2015). Tracing the water–energy–food nexus: description, theory and practice. *Geography Compass* **9**, 445–460.

Liu, J., Yang, H., Cudennec, C., Gain, A., Hoff, H., Lawford, R., Qi, J., Strasser, L. d., Yillia, P., and Zheng, C. (2017). Challenges in operationalizing the water–energy–food nexus. *Hydrological Sciences Journal* **62**, 1714–1720.

Mabhaudhi, T., Mpandeli, S., Madhlopa, A., Modi, A. T., Backeberg, G., and Nhamo, L. (2016). Southern Africa's water–energy nexus: Towards regional integration and development. *Water* **8**, 235.

Mabhaudhi, T., Mpandeli, S., Nhamo, L., Chimonyo, V. G., Nhemachena, C., Senzanje, A., Naidoo, D., and Modi, A. T. (2018b). Prospects for improving irrigated agriculture in southern Africa: Linking water, energy and food. *Water* **10**, 1881, 1–16, doi:10.3390/w10121881.

Mabhaudhi, T., Mpandeli, S., Nhamo, L., Chimonyo, V., Nhemachena, C., Senzanje, A., Naidoo, D., and Modi, A. (2018a). Prospects for improving irrigated agriculture in southern Africa: Linking water, energy and food. *Water* **10**, 1881.

Machalaba, C., Romanelli, C., Stoett, P., Baum, S. E., Bouley, T. A., Daszak, P., and Karesh, W. B. (2015). Climate change and health: Transcending silos to find solutions. *Annals of Global Health* **81**(3). doi:10.1016/j.aogh.2015.08.002.

McGrane, S. J., Acuto, M., Artioli, F., Chen, P. Y., Comber, R., Cottee, J., Farr-Wharton, G., Green, N., Helfgott, A., and Larcom, S. (2018). Scaling the nexus: Towards integrated frameworks for analysing water, energy and food. *The Geographical Journal* **185**(4), 419–431.

Mohtar, R. H., and Daher, B. (2016). Water-energy-food nexus framework for facilitating multi-stakeholder dialogue. *Water International* **41**, 655–661.

Mpandeli, S., Naidoo, D., Mabhaudhi, T., Nhemachena, C., Nhamo, L., Liphadzi, S., Hlahla, S., and Modi, A. (2018). Climate change adaptation through the water-energy-food nexus in Southern Africa. *International Journal of Environmental Research and Public Health* **15**, 2306, 1–19. doi:10.3390/ijerph15102306.

NEPAD (2014). Implementation Strategy and Roadmap to Achieve the 2025 Vision on CAADP. Operationalizing the 2014 Malabo Declaration on Accelerated African Agricultural Growth and Transformation for Shared Prosperity and Improved Livelihood. NEPAD, Addis Ababa, Ethiopia.

Nhamo, L. (2015). *Trends and Outlook: Agricultural Water Management in Southern Africa. SADC AgWater Profiles.* International Water Management Institute (IWMI), Southern Africa Regional Office, Pretoria, South Africa.

Nhamo, L., Mabhaudhi, T., and Modi, A. (2019a). Preparedness or repeated short-term relief aid? Building drought resilience through early warning in southern Africa. *Water SA* **45**, 20.

Nhamo, L., Mabhaudhi, T., Mpandeli, S., Nhemachena, C., Sobratee, N., Naidoo, D., Liphadz, S., and Modi, A. (2019b). Sustainability indicators and indices for the water-energy-food nexus for performance assessment: WEF nexus in practice–South Africa case study. *Preprints*, 2019050359. doi:10.20944/preprints201905.0359.v1.

Nhamo, L., Matchaya, G., Mabhaudhi, T., Nhlengethwa, S., Nhemachena, C., and Mpandeli, S. (2019c). Cereal production trends under climate change: Impacts and adaptation strategies in southern Africa. *Agriculture* **9**, 16.

Nhamo, L., Ndlela, B., Nhemachena, C., Mabhaudhi, T., Mpandeli, S., and Matchaya, G. (2018). The water-energy-food nexus: Climate risks and opportunities in Southern Africa. *Water* **10**, 18.

Niang, I., Osman-Elasha, B., Githeko, A., Yanda, P. Z., Medany, M., Vogel, A., Boko, M., Tabo, R., and Nyong, A. (2008). Africa Climate Change 2007: Impacts, Adaptation and Vulnerability: Contribution of Working Group II to the Fourth Assessment Report of the Intergovernmental Panel on Climate Change. Cambridge University Press, Cambridge, UK.

Niang, I., Ruppel, O. C., Abdrabo, M. A., Essel, A., Lennard, C., Padgham, J., and Urquhart, P. (2014). In: *Climate Change 2014: Impacts, Adaptation, and Vulnerability. Part B: Regional Aspects. Contribution of Working Group II to the Fifth Assessment Report of the Intergovernmental Panel on Climate Change*, Cambridge University Press, Cambridge, UK and New York.

Nyikahadzoi, K., Mhlanga, W., Madzudzo, E., Tendaupenyu, I., and Silwimba, E. (2017). Dynamics of transboundary governance and management of small scale fisheries on Lake Kariba: Implications for sustainable use. *International Journal of Environmental Studies* **74**, 458–470.

Nzuma, J. M., Waithaka, M., Mulwa, R. M., Kyotalimye, M., and Nelson, G. (2010). Strategies for adapting to climate change in sub-Saharan Africa: A review of data sources, poverty reduction strategy programs (PRSP) and national adaptation plans for agriculture (NAPAs) in ASARECA member countries. IFPRI Discussion Paper 01013, International Food Policy Research Institute (IFPRI), Washington, DC.

Parry, M., Parry, M. L., Canziani, O., Palutikof, J., Van der Linden, P., and Hanson, C. (2007). *Climate Change 2007-Impacts, Adaptation and Vulnerability: Working Group II Contribution to the Fourth Assessment Report of the IPCC*, Cambridge University Press, Cambridge, UK.

Rees, J. (2013). Geography and the nexus: Presidential address and record of the Royal Geographical Society (with IBG) AGM 2013. *The Geographical Journal* **179**, 279–282.

SADC (1996). Protocol on Energy in the Southern African Development Community (SADC). Southern African Development Community (SADC), Gaborone, Botswana.

SADC (2000). Revised Protocol on Shared Watercourses in the Southern African Development Community. Southern African Development Community (SADC), Gaborone, Botswana.

SADC (2011). *The Consolidated Treaty of the Southern Africa Development Community (SADC)*. Southern Africa Development Community (SADC), Gaborone, Botswana.

SADC (2012). *SADC Regional Infrastructure Development Master Plan: Executive Summary*. Southern Africa Development Community (SADC), Gaborone, Botswana, p. 40.

SADC (2014a). *SADC Regional Agricultural Policy*. Southern African Development Community (SADC), Gaborone, Botswana.

SADC (2014b). *SADC Regional Agricultural Policy*. Southern African Development Community (SADC), Gaborone, Botswana.

SADC (2015). *Regional Strategic Action Plan on Integrated Water Resources Development and Management (2016–2020) RSAP IV*. Southern Africa Development Community (SADC), Gaborone, Botswana.

SADC (2016). *Regional Strategic Action Plan on Integrated Water Resources Development and Management Phase IV*. Southern African Development Community (SADC), Gaborone, Botswana.

Schreiner, B., and Baleta, H. (2015). Broadening the lens: A regional perspective on water, food and energy integration in SADC. *Aquatic Procedia* **5**, 90–103.

Scott, C. A., Kurian, M., and Wescoat Jr, J. L. (2015). The water-energy-food nexus: Enhancing adaptive capacity to complex global challenges. In *Governing the Nexus* (M. Kurian and R. Ardakanian, eds.), pp. 15–38. Springer, Dresden.

Stiles, G., and Murove, C. (2015). *SADC Renewable Energy and Energy Efficiency Status Report*. Renewable Energy Policy Network for the 21st Century (REN21), Paris.

Terrapon-Pfaff, J., Ortiz, W., Dienst, C., and Gröne, M.-C. (2018). Energising the WEF nexus to enhance sustainable development at local level. *Journal of Environmental Management* **223**, 409–416.

The World Bank (2010). *The Zambezi River Basin A Multi-Sector Investment Opportunities Analysis—State of the Basin*. The World Bank, Washington, DC.

Thornton, P. K., Ericksen, P. J., Herrero, M., and Challinor, A. J. (2014). Climate variability and vulnerability to climate change: A review. *Global Change Biology* **20**, 3313–3328.

van Koppen, B., Lacombe, G., and Mwendera, E. (2015). *Trends and Outlook: Agricultural Water Management in Southern Africa: Synthesis Report*. International Water Management Institute (IWMI), Colombo, Sri Lanka.

Waldinger, M. (2015). *The Effects of Climate Change on Internal and International Migration: Implications for Developing Countries*. Centre for Climate Change Economics and Policy, London.

Wlokas, H. (2008). The impacts of climate change on food security and health in Southern Africa. *Journal of Energy in Southern Africa* **19**, 12–20.

8 Green Technology to Mitigate Global Water, Energy, and Environment

Md. Faruque Hossain

CONTENTS

8.1 INTRODUCTION

Plants take carbon dioxide (CO_2) and release oxygen (O_2) through photosynthesis, and the process maintains balance in the global environment [1,7,9]. Therefore, plants play a central role in environmental equilibrium. However, plants, despite their critical role, contribute significantly to form aerosol into the air. For plants to grow, their bodies need water to facilitate biochemical metabolism. The plants rely on cohesion-tension mechanisms to absorb the ground water in the soil via the roots [12,13]. The process of osmosis facilitates the movement of water via the xylems to the leaves. However, plants only use 0.5% of the absorbed water in the metabolism process [36,37]. They release the remainder 95.5% water into the air via stomatal cells in a process referred as transpiration [2,36]. This process is the largest case of groundwater loss, and it contributes to weather change because the water vapor contributes significantly to form aerosol into the air.

Thus, the current research proposes a technology that can be used to eliminate or reduce water loss by diverting the mechanism involved in transpiration by collecting the water vapor rather than allow its entry into the air. The proposed technology would transform the water vapor into both clean energy and portable water. Placing a static electricity creator plastic tank is a solution that has been proposed to trap the water vapor by using the force of static electricity to attract the water vapor. The logic behind the technology is the realization that water vapor contains both negative and positive charges as well as electrons. If there is to be an electrical force with a positive charge and the water molecules have an aggregate negative charge, the two forces end up pulling each other [10,11,17]. The positively charged side would pull the water vapor and direct it into a tank in which it can be treated for domestic use. Part of the collected water is to be subjected to an electrolysis process for clean energy (hydrogen) production that will release O_2 as a by-product. This by-product could be useful in balancing the environment when released into the air.

The conducted calculations show that five standard oak trees can sufficiently meet the total energy and water needs of a small family in a year. With the level of ground water strata lowering rapidly and the global warming and global energy continuing to leave the Earth in a precarious situation, an immediate solution to these needs becomes necessary. If followed, the proposed solution has the potential to solve the global energy, water, and environmental crises threatening Earth's survival.

8.2 MATERIAL, METHODS, AND SIMULATION

8.2.1 STATIC ELECTRIC FORCE GENERATION

This study proposes a model of creating *Hossain Static Electric Force/Field (HSEF=ƒ)* that will facilitate the capture of water from air and its consequence release by plants' stomatal cells during the day. The mechanism in the proposed model uses the friction from the insulator in contact with the plastic tank and which pulls the water vapor down into the plastic tank [18,21]. When attempting to incorporate the *HSEF* into the plastic tank, performing abelian local symmetries calculation using the MATLAB® software enables one to consider the gauge field symmetry as well as the Goldstone scalar in relation to the longitudinal mode that the vector assumes [24,28]. Therefore, each particle T^α of the local symmetry that is spontaneously broken has a corresponding gauge field of $A_\mu^\alpha(x)$ with *HSEF* beginning to function at local U(1) phase symmetry [19,23,26]. Thus, the model constitutes of a complex scalar field $\Phi(x)$ with a static electric charge of q integrated with the electromotive (EM) field $A^\mu(x)$. The expression of the model includes ƒ:

$$= -\frac{1}{4}F_{\mu\nu}F^{\mu\nu} + D_\mu\Phi^* D^\mu\Phi - V\left(\Phi^*\Phi\right) \tag{8.1}$$

where

$$D_\mu\Phi(x) = \partial_\mu\Phi(x) + iqA_\mu(x)\Phi(x)$$

$$D_\mu\Phi^*(x) = \partial_\mu\Phi^*(x) - iqA_\mu(x)\Phi^*(x) \tag{8.2}$$

And

$$V\left(\Phi^*\Phi\right) = \frac{\lambda}{2}\left(\Phi^*\Phi\right)^2 + m^2\left(\Phi^*\Phi\right) \tag{8.3}$$

Assume that $\lambda > 0$ but $m^2 < 0$, which implies that $\Phi = 0$ represents the local maximum that the scalar potential has with the minima forming a degenerate circle $\Phi = \frac{v}{\sqrt{2}} * e^{i\Theta}$, in which:

$$v = \sqrt{\frac{-2m^2}{\lambda}}, \text{ any real } \theta \tag{8.4}$$

As a result, the scalar field Φ leads to a non-zero vacuum expectation value $\Phi \neq 0$ that contributes in creating U(1) symmetry that the magnetic field adopts [3,4,5]. The consequent breakdown causes a massless Goldstone scalar that stems from the complex field $\Phi(x)$ phase. However, in case of the local U(1) symmetry, $\Phi(x)$ not only covers the expectation value Φ because the x-dependent variable in the dynamic $\Phi(x)$ field.

In confirming the mechanism applicable in static electricity force, this study employs polar coordinates in the applicable scalar field space. Therefore:

$$\Phi(x) = \frac{1}{\sqrt{2}} \Phi_r(x) * e^{i\Theta(x)}, \text{ real } \Phi_r(x) > 0, \text{ real } \Phi(x) \tag{8.5}$$

The redefinition of the field remains singular as $\Phi(x) = 0$. Thus, this study did not use it in case of any theory with $\Phi \neq 0$. However, it would suit in theories that are spontaneously broken because $\Phi\langle x \rangle \neq 0$ is expected to be observed everywhere. In relation to real fields of $\Theta(x)$ and $\phi_r(x)$, the radial field ϕ_r is the only factor influencing the scalar potential:

$$V(\phi) = \frac{\lambda}{8}\left(\phi_r^2 - v^2\right)^2 + \text{const,} \tag{8.6}$$

Or in regard to the radial field that its Vacuum Expectation Value (VEV) causes to shift, $\Phi_r(x) = v + \sigma(x)$,

$$\phi_r^2 - v^2 = (v + \sigma)^2 - v^2 = 2v\sigma + \sigma^2 \tag{8.7}$$

$$V = \frac{\lambda}{8}\left(2v\sigma - \sigma^2\right)^2 = \frac{\lambda v^2}{2} * \sigma^2 + \frac{\lambda v}{2} * \sigma^3 + \frac{\lambda}{8} * \sigma^4 \tag{8.8}$$

Concurrently, the covariant derivative $D_\mu\phi$ comprises:

$$D_\mu\phi = \frac{1}{\sqrt{2}}\left(\partial_\mu\left(\phi_r e^{i\Theta}\right) + iqA_\mu * \phi_r e^{i\Theta}\right) = \frac{e^{i\Theta}}{\sqrt{2}}\left(\partial_\mu\phi_r + \phi_r * i\partial_\mu\Theta + \phi_r * iqA_\mu\right) \tag{8.9}$$

$$\left|D_\mu\phi\right|^2 = \frac{1}{2}\left|\partial_\mu\phi_r + \phi_r * i\partial_\mu\Theta + \phi_r * iqA_\mu\right|^2$$

$$= \frac{1}{2}\left(\partial_\mu\phi_r\right) + \frac{\phi_r^2}{2} * \left(\partial_\mu\Theta qA_\mu\right)^2 \tag{8.10}$$

$$= \frac{1}{2}\left(\partial_\mu\sigma\right)^2 + \frac{(v+\sigma)^2}{2} * \left(\partial_\mu\Theta + qA_\mu\right)^2$$

In total,

$$\mathfrak{H} = \frac{1}{2}\left(\partial_\mu \sigma\right)^2 - v\left(\sigma\right) - \frac{1}{4}F_{\mu v}F^{\mu v} + \frac{\left(v+\sigma\right)^2}{2}*\left(\partial_\mu \Theta + qA_\mu\right)^2 \qquad (8.11)$$

In confirming the creation of the static electric force ($\mathfrak{H}_{sef}$) and inclusion in the electric field attributes of the HSEF, the force is expanded in powers of the various powers together with their derivatives with emphasis being given to the quadratic part that illustrates the free particles,

$$\mathfrak{H}_{sef} = \frac{1}{2}\left(\partial_\mu \sigma\right)^2 - \frac{\lambda v^2}{2}*\sigma^2 - \frac{1}{4}F_{\mu v}F^{\mu v} + \frac{v^2}{2}*\left(qA_\mu + \partial_\mu \Theta\right)^2 \qquad (8.12)$$

The *HSEF* ($\mathfrak{H}_{free}$) function provided is expected to propose an actual scalar particle with positive mass$^2 = \lambda v^2$ that involves both $A_\mu(x)$ and $\Theta(x)$ fields for initiating the creation of a remarkable static electricity force in the plastic tank's electric field as illustrated in Figure 8.1.

8.2.2   IN-SITE WATER TREATMENT

The water that the plastic tank collects is the liquid form of the water vapor. No sedimentation, chlorination, and coagulation will be needed to clean it. Ultraviolet (UV) application/mixing physics and filtration would be sufficient in treating the water for it to meet the set US National Primary Drinking Water Standard code [14,34,35]. That constitutes the simplest way of treating water involving *Solar Disinfection* (SODIS) system in which one fills a transparent container with water and exposes it to full sunlight for a few hours. Once the water temperature hits 50°C when subject to a UV radiation of approximately 320 nm, this accelerates the inactivation process

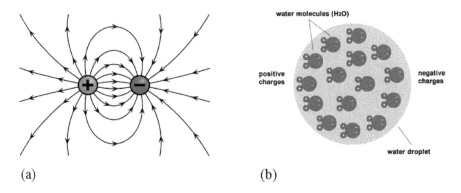

(a)   (b)

FIGURE 8.1   (a) The creation of a static electricity force, and (b) the mechanism involved in converting static energy into an electromotive force of negative and positive charges, which gather the "static" electricity together to pull the water molecules down.

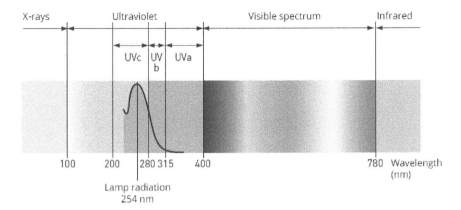

FIGURE 8.2   The application of photo-physics radiation in purifying water that illustrates that once one applies ultraviolet (UV) radiation of 320 nm into water, it begins to disinfect the microorganisms the moment the temperature hits 50°C.

that immediately causes thorough bacteriological disinfection. The treated water can then be used for domestic purposes (Figure 8.2).

8.2.3   CLEAN ENERGY PRODUCTION

Some amount of the water will be used in generating clean energy through *electrolysis*, a process of converting water into hydrogen energy. Rather than employ the traditional approach, this study proposes a direct water electrolysis system that is based on a new, integrated, and monolithic photo-electrochemical (PEC)/photovoltaic (PV) device that Figure 8.4 illustrates. The device is similar to the GaInP$_2$/GaAs *p/n*, *p/n* tandem cell device. In this device, the solid-state tandem cell is made up of a GaAs bottom cell that is linked to a GaInP$_2$ top cell via the tunnel diode linkage. Whereas, the top *p/n* GaInP$_2$ junction having a band gap of 1.83 eV tends to be designed to absorb the visible section of the solar spectrum. In case of the bottom *p/n* GaInP$_2$ junction having a band gap of approximately 1.42 eV, the junction absorbs the almost infrared section of the spectrum that is transmitted via the top junction for the excited radiant energy to conduct electrolysis as Figure 8.3 shows.

The optimal theoretical solar-to-electrical energy efficiency for the band gaps that are presently combined is to be calculated if the standard solid-state tandem cell in implementing the PEC Schottky-type junction is to be achieved at the top of the *p/n* junction. The consequence would be a PEC device that is voltage-biased and has an integrated PV device. Under illumination, the electrons move to the illuminated surfaces and the holes toward the ohmic contact to splitting the water [6,8,16]. On illumination, the water splits. For this to happen, light is the only input made to the PEC device. The *p*-GaInP$_2$/GaAs that the study used were developed using atmospheric-pressure organometallic vapor-phase epitaxy. Such a method involves the top layer of the epitaxially cell, *p*-Ga$_{0.52}$In$_{0.48}$P (as *p*-GaInP$_2$) that is 4.0 ± 0.5 µm thick. It is connected in series through a low-resistivity and cell-in tunnel junction

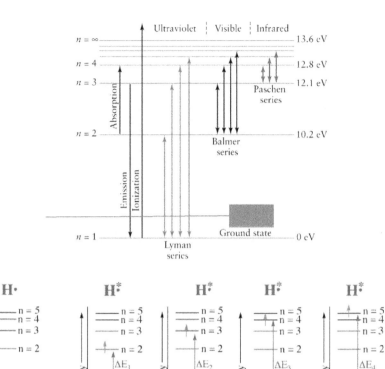

FIGURE 8.3   The clarification of electron-state hydrogen energy is achieved using the radiation emission of photon energy (ultraviolet [UV] light) having a band gap of 1.42 eV for the clarification for photo-electrolysis to achieve hydrogen that is in an excited state.

(TJ) to a GaAs *p/n* bottom cell achieved on a GaAs substrate for confirmation of the achievement of optimal results. As a result, the standard electrochemical and chemical procedures are employed [20] with a platinum catalyst being used to coat the surface of the samples [21]. The generation of illumination on which the photo-electrolysis relies is to involve a fiber-optic illuminator that has a 150 W tungsten-halogen lamp. Measurements of a light irradiance levels at the surface of the surface involves mounting a calibrated PV-tandem cell into the electrode holder contained inside the cell as it applies when photoelectrodes are placed under the light irradiance measured before [22].

In case of the PEC/PV configuration, one has to supply the GaAs bottom cell with sufficient voltage for the configuration to function appropriately. The voltage will need to overcome energetic mismatch that the band edges of the GaInP$_2$ have with the water redox reactions [27,29,33]. Furthermore, any additional voltage that is necessary to

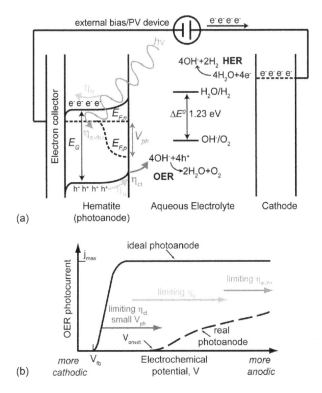

(a)

(b)

FIGURE 8.4 (a) Schematic diagram of the monolithic bias photoelectrochemical/ photovoltaic (PEC/PV) device, (b) Diagram of the idealized photoanode energy level for monolithic PEC/PV photoelectrolysis device.

overcome overvoltage losses that arise in hydrogen (H_2) and O_2 evolution reactions has to be factored also (Figure 8.4). In total, the photovoltage output approaches the thermodynamics associated with natural water splitting (1.23 V). These include polarization losses μ_a and μ_c that characterize anodic and cathodic processes, respectively. Figure 8.4b demonstrates an idealized energy-level diagram detailing the photolytic splitting of water using the device. In the illustrated process there are two photons together with one separate electron-hole pair. Initially, the light that is incident on the PEC/PV configuration gains entry to the broad band gap p-$GaInP_2$ layer that absorbs the more energetic photons. The process excites the electron hole and generates photovoltage V_{ph1}, so that the less energetic photons gain entry via the $GaInP_2$ and the GaAs bottom p/n junction absorbs them to generate photovoltage V_{ph2} (Figure 8.4). One group of holes and electrons get recombined at the junction of the tunnel. Once the resultant photovoltage $V_{ph} = V_{ph1} + V_{ph2}$ exceeds the one that photo-electrolysis require for this specific cell configuration, that will prompt it to drive the water at the semiconductor electrode and water at the counter electrode requiring only two photons to generate an electron in the external circuit, while only four electrons are to generate one H_2 molecule, which reflects clean energy generation.

8.3 RESULTS AND DISCUSSION

8.3.1 ELECTROSTATIC FORCE ANALYSIS

In an attempt to establish the proliferation of electric force around the plastic tank mathematically to tug the water down, the dynamic photon proliferation is solved through the integration of HSEF electric field created. Therefore, the local U(1) gauge invariant allowed the addition of mass term for gauge particles within $\varnothing' \to e^{i\alpha(x)}\varnothing$. A covariant derivative using a special transformation rule can explain it in detail for the scalar field as expressed in [37,12,13]:

$$\partial_\mu \to D_\mu = \partial_\mu = ieA_\mu \text{ [covariant derivatives]}$$

$$A'_\mu = A_\mu + \frac{1}{e}\partial_\mu\alpha \text{ [}A_\mu \text{ derivatives]} \tag{8.13}$$

Whereby local U(1) gauge invariant *HSEF* suited to the complex scalar field is represented by:

$$\mathfrak{H} = \left(D^\mu\right)^\dagger \left(D_\mu\varnothing\right) - \frac{1}{4}F_{\mu\nu}F^{\mu\nu} - V\left(\varnothing\right) \tag{8.14}$$

Notably, the function $\frac{1}{4}F_{\mu\nu}F^{\mu\nu}$ represents the kinetic term for the gauge field, that is, heating photon, and $V\left(\varnothing\right)$ represents the extra term contained in the HSEF. Such means $V(\varnothing^*\varnothing) = \mu^2(\varnothing^*\varnothing) + \lambda\ (\varnothing^*\varnothing)^2$.

Thus, the *HSEF* ($\mathfrak{H}$), when subject to perturbations in the quantum field, start with massive scalar particles ϕ_1 and ϕ_2 together with a mass μ. The consequent situation $\mu^2 < 0$ involves countless quantum, and every one meets the condition $\phi_1^2 + \phi_2^2 = -\mu^2 / \lambda = v^2$ and the $\mathfrak{H}$ via the covariant derivatives that also use the shifted fields η and ξ that the quantum field defines as $\phi_0 = \frac{1}{\sqrt{2}}\left[(v+\eta) + i\xi\right]$.

Kinetic term is represented as $\mathfrak{H}_{kin}(\eta,\xi) = (D^\mu\phi)^\dagger(D^\mu\phi)$

$$= \left(\partial^\mu + ieA^\mu\right)\phi^*\left(\partial_\mu - ieA_\mu\right)\phi \tag{8.15}$$

Potential term includes $V\left(\eta,\xi\right) = \lambda v^2\eta^2$, and it applies up to the second order in various fields. The complete *HSEF* can, therefore, be presented as:

$$\mathfrak{H}_{kin}(\eta,\xi) = \frac{1}{2}\left(\partial_\mu\eta\right)^2 - \lambda v^2\eta^2 + \frac{1}{2}\left(\partial_\mu\xi\right)^2 - \frac{1}{4}F_{\mu\nu}F^{\mu\nu} + \frac{1}{2}e^2v^2A_\mu^2$$

$$- evA_\mu\left(\partial^\mu\xi\right) + \text{int. terms} \tag{8.16}$$

In this case, massive is represented as η, massless as ξ (as previously done) and a mass term for quantum A_μ, which is fixed in number up to $\partial_\mu\alpha$ as Eq. (8.14) demonstrates. A_μ and ϕ vary simultaneously, and therefore, it is possible to redefine them to incorporate the heating photon particle spectrum that falls in the quantum field through the expression:

$$\mathfrak{H}_{scalar} = \left(D^\mu\phi\right)^\dagger \left(D^\mu\phi\right) - V\left(\phi^\dagger\phi\right)$$

$$= \left(\partial^\mu + ieA^\mu\right)\frac{1}{\sqrt{2}}(v+h)\left(\partial_\mu - ieA_\mu\right)\frac{1}{\sqrt{2}}(v+h) - V\left(\phi^\dagger\phi\right)$$

$$= \frac{1}{2}\left(\partial_\mu h\right)^2 + \frac{1}{2}e^2A_\mu^2(v+h)^2 - \lambda v^2h^2 - \lambda vh^3 - \frac{1}{4}\lambda h^4 + \frac{1}{4}\lambda h^4 \qquad (8.17)$$

Therefore, the expansion term in the $\hat{g}$ linked to the scalar field suggests that *HSEF* electric field is ready to start the propagation of static electricity to create a quantum field capable of tugging the water down.

Confirming the tugging down of water requires calculations of isotropic spread of movement on the differential cone taking into account angle θ and remaining in the range between θ and $\theta + d\theta$ constitutes $\frac{1}{2}\sin\theta d\theta$ and with a differential static electric force density recorded at energy $\in$ and angle θ as shown:

$$dn = \frac{1}{2}n(\in)\sin\theta d \in d\theta \qquad (8.18)$$

Therefore, the calculation of the function that the high static electricity force plays takes into account the directional form of $c\,(1 - \cos\theta)$, calculating the absorption level of water vapor for each unit as:

$$\frac{d\tau_{abs}}{dx} = \int\int \frac{1}{2}\sigma n\,(\in)(1-\cos\theta)\sin\theta d \in d\theta. \qquad (8.19)$$

In an effort to modify the functions to realize an integration rather than s rather than θ by (8.3) and (8.5), the calculation of determining the accuracy are as follows:

$$\frac{d\tau_{abs}}{dx} = \pi r_0^2 \left(\frac{m^2c^4}{E}\right)^2 \int\limits_{\frac{m^2c^4}{E}}^{\infty} \in^{-2} n(\in)\,\bar{\phi}[s_0(\in)]de, \qquad (8.20)$$

in which,

$$\bar{\varphi}[s_0(\in)] = \int\limits_{1}^{s_0(\in)} s\bar{\sigma}\,(s)ds, \quad \bar{\sigma}(s) = \frac{2\sigma(s)}{\pi r_0^2}. \qquad (8.21)$$

The results obtained emerge as a dimension variable $\bar{\varphi}$ together with dimension-less cross section $\bar{\sigma}$. Calculations of the variable $\bar{\varphi}[s_0]$ are made based on a detailed graphical frame for $1 < s_0 < 10$. The functional asymptotic calculation made for $\bar{\varphi}$ had to be reliable whereby $s_0 - 1 \ll 1$ and $s_0 \gg 1$ is expressed as:

$$\bar{\varphi}[s_0] = \frac{1+\beta_0^2}{1-\beta_0^2}\ln\omega_0 - \beta_0^2\ln\omega_0 - \ln^2\omega_0 - \frac{4\beta_0}{1-\beta_0^2} + 2\beta_0 + 4\,\ln\omega_0\ln\left(\omega_0+1\right) - L\left(\omega_0\right),$$

where

$$\beta_0^2 = \frac{1-1}{s_0}, \omega_0 = \frac{(1+\beta_0)}{(1-\beta_0)},$$

(8.22)

$$L(\omega_0) = \int_1^{\omega_0} \omega^{-1} \ln (\omega+1) d\omega.$$

It would be appropriate to write the last integral as

$$(\omega + 1) = \omega\left(\frac{1+1}{\omega}\right), L(\omega_0) = \frac{1}{2}\ln^2\omega_0 + L'(\omega_0),$$

where

$$L'(\omega_0) = \int_1^{\omega_0} \omega^{-1} \ln\left(1+\frac{1}{\omega}\right) d\omega,$$

(8.23)

$$= \frac{\pi^2}{12} - \sum_{n=1}^{\infty} (-1)^n n^{-2} \omega_0^{-n}.$$

Ultimately, the expression of the tug down water using static electricity force can be detailed readily by implementing the calculations of $\overline{\varphi}[s_0]$ to confirm the expected s_0 value to capture water vapor. Therefore, this study uses the corrective functional asymptotic formulas as provided below:

$$\overline{\varphi}[s_0] = 2s_0(\ln 4s_0 - 2) + \ln 4s_0(\ln 4s_0 - 2) - \frac{(\pi^2-9)}{3} + s_0^{-1}\left(\ln 4s_0 + \frac{9}{8}\right) + ... \ (s_0 \gg 1);$$

(8.24)

$$\overline{\varphi}[s_0] = \left(\frac{2}{3}\right)(S_0-1)^{\frac{3}{2}} + \left(\frac{5}{3}\right)(S_0-1)^{\frac{5}{2}} - \left(\frac{1507}{420}\right)(S_0-1)^{\frac{7}{2}} + ... \ (s_0 - 1 \ll 1). \quad (8.25)$$

The function $\frac{\phi[s_0]}{(s_0-1)}$ is therefore illustrated as $1 < s_0 < 10$. In case of the larger s_0, it has a natural logarithmic that is s_0 for confirming the tug down of 100% water vapor that the HSEF direct to the plastic tank.

In typical daily life, every person requires an average of 100 gallons of water every day. Therefore, for 1 year, $100_{gallons}$/Day/Person $\times 4_{persons} \times 365_{days}$, which amounts to 146,000 gallons of water that a family of four would require annually. Normally, a standard oak tree transpires an approximately 151,000 liters or 40,000 gallons of water annually. This means that only four standard oak trees would be needed to satisfy the family's water demand if the HSEF manages to tug down 100% of the water vapor. If the family has 6 standard oak trees, it can rely four to meet its water domestic water demands and the others for generating clean energy for the home through the electrolysis process.

8.4   ELECTROLYSIS FOR HYDROGEN ENERGY PRODUCTION

In the current model, it is easy to conduct water-splitting reactions for H_2 production because (a) the H_2 evolution reaction tends to have the lowest over-voltage losses, which lowers the need to have a catalyst and, therefore, facilitates an optimized counter electrode for use in more sophisticated O_2 evolution reactions, and (b) under illumination, which requires the semiconductor surface to be protected cathodically [30,31,32]. Figure 8.5 illustrates photocurrent-voltage curves applicable to p-GaInP$_2$(Pt)/TJ/GaAs and p-GaInP$_2$(Pt) electrodes assessed in a two-electrode configuration. In the dark, the open circuit voltage is expected to be −0.75 V and ~−0.64 for p-GaInP$_2$(Pt)/TJ/GaAs and p-GaInP$_2$(Pt) electrodes respectively. Notably, the dark reduction current is to remain in the microampere range for the two electrodes. The p-GaInP$_2$ (Pt) while under illumination is to be started to produce hydrogen at

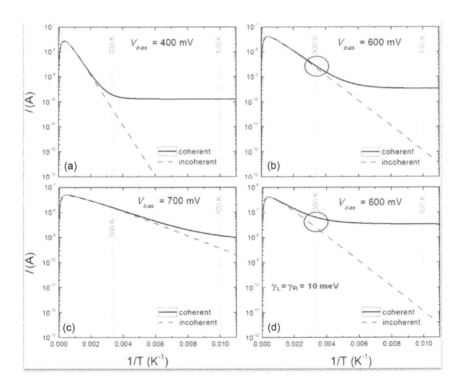

FIGURE 8.5   Electrical current via a single-level molecular junction as Eq. (8.18) (broadened molecular level) and Eq. (8.19) (zero-width molecular level) express. The parameters that the calculations use include $\varepsilon = 0.4$ eV, $\eta = 0.5$, $V_G = 0$, and $\gamma_L = \gamma_R = 1$ meV for (a–c) and $\gamma_L = \gamma_R = 10$ meV for (d). These parameters are typical values in the molecular junction of the current work.

a 500 mV negative with 0 V bias. Such implies that additional external voltage will have to be provided for the semiconductor to manage splitting the water. Whereas, p-GaInP$_2$ (Pt)/TJ/GaAs electrode exhibits an open circuit voltage of ~0.55 V when under illumination. That suggests that the GaAs cell generates the extra voltage. H$_2$ evolution begins directly and at 400 mV positive of a short circuit. The density of photocurrent is to be met at a limiting value of 120 mA/cm^2 and at ~0.15 V. This would remain constant with rising bias. Numerous gas bubbles are to be seen at the semiconductor surface. Because the gas bubbles are able to reach a size that they can sufficiently act as miniature lenses to pit the semiconductor electrode, that means that 0.01 M of the surfactant Triton X-100 is chosen as the solution of facilitating the formation of smaller bubbles, which leaves the sample surface faster. The less saturated photocurrent applicable in p-GaInP$_2$ (Pt)/TJ/GaAs electrode in comparison to the p-GaInP$_2$ electrode is considered in showing that the p/n GaAs bottom cell constitutes the current-limiting junction (Figure 8.5).

The collection and analysis of products of photoelectrolysis will be conducted by mass spectrometer. The calculations of the efficiency of generating H$_2$ will follow the equation: Efficiency = (power out)/(power in). Hereby, the input power constitutes the incident light intensity of approximately 1190 mW/cm^2 (Figure 8.6a). In case of the output power, taking the assumption of 100% efficiency in photocurrent electrolysis, the H$_2$ generation photocurrent of 120 mA/cm^2 get multiplied with 1.23 V that constitutes the ideal fuel cell limit achieved at 25°C to arraign the highest efficiency in H$_2$ energy generation; 25°C is the lowest heating value of hydrogen (Figure 8.6b).

The calculated estimate shows that a gallon of water can yield 0.42 kg of hydrogen. In the proposed PEM electrolyzers, the ideal figure would be 44.5 kWh/kg-H$_2$. Thus, the yields amount to 16.7 kWh. Notably, a kilowatt hour constitutes 3600 kilojoule and the enthalpy of forming hydrogen energy derived from liquid water and at 25°C amounts to −286 kJ/mol, while a gallon of water constitutes 3785 mL. Taking into account that water ought to be 18 mL/mol, which is approximately 210 mol of

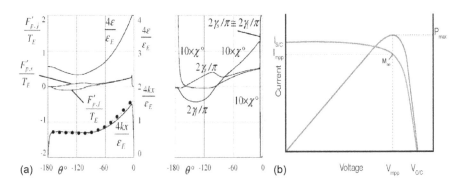

FIGURE 8.6   (a) Current-voltage attributes for curve 1 (p-GaInP$_2$(Pt)/TJ/GaAs) and curve 2 (p-GaInP$_2$) electrodes in 3MH 2SO$_4$ when placed under white light illumination. (b) A photocurrent time profile recorded at short circuit for photoelectrochemical/photovoltaic (PEC/PV) tandem cell in 3MH 2SO$_4$ using 0.01 M Triton X-100 when placed under tungsten-halogen white light illumination. The recorded efficiency of 5120 mA/cm^2 31.23 V 3100/1190 mW/cm^2 at current density were observed.

water, the calculation $286 \times 210.3/3600$ will imply that 116.7-kilowatt-hours energy is generated. On average a small family of four requires 30 kWh daily. As a result, two gallons of water would be sufficient to meet their daily energy demands if done on daily basis. Thus, the amount water vapor that a small oak tree can generate would be sufficient if electrolysis applies.

8.5   CONCLUSION

Plants play the major role in balancing the global environmental equilibrium. Nevertheless, the process of transpiration uses only a mere portion of the ground water and releases rest of the water into the air. Simply, the hydraulic conductivity that the soil has together with the scale of pressure gradient via the soil affects the large amount of water that gets to the plant leaves from the roots. The cohesive attribute of the water enables tension to pass through leaf cells to both the leaf and stem xylem. From this, a momentary negative pressure results on the pulling of water from the roots and up the xylem. Surprisingly, plants' metabolism only uses 0.5% and the remaining portion of 99.5% evaporates through transpiration through the stomatal cell. Ultimately, there is continuous water flowing through plants leading to loss of considerable amounts of ground water that creates significant roles in global potable water crises throughout the world. As a way of mitigating this issue, the transpiration mechanism is suggested to be useful in transforming and converting the transpiration water vapor into clean energy and clean water and consequently mitigate global potable water crisis. The adoption of this technology promises to be a revolutionary field of science that humans can use to tackle the global potable water and global energy crisis in the hope of establishing the best-balanced Earth and a humane civilization.

ACKNOWLEDGMENTS

Green Globe Technology offered support in preparing this research under grant RD-02017-06 aiming to build a better environment. The author came up with all the findings, assumptions, conclusions, and prediction in the article and there was no conflict of interest in choosing to publish the research in a journal.

REFERENCES

1. Alvar R. Garrigues, Li Yuan, Lejia Wang, Eduardo R. Mucciolo, Damien Thompon, Enrique del Barco, and Christian A. Nijhuis. A single-level tunnel model to account for electrical transport through single molecule- and self-assembled monolayer-based junctions, *Scientific Reports* 6, 2016.
2. Andreas Reinhard. Strongly correlated photons on a chip. *Nature Photonics* 6, 93–96, December 18, 2011.
3. J. S. Douglas, H. Habibian, C.-L. Hung, A. V. Gorshkov, H. J. Kimble, and D. E. Chang. Quantum many-body models with cold atoms coupled to photonic crystals. *Nature Photonics* 9, 326–331, 2015.
4. G. Baur, K. Hencken, and D. Trautmann. Revisiting unitarity corrections for electromagnetic processes in collisions of relativistic nuclei. *Physics Report* 453, 1, 2007.

5. G. Baur, K. Hencken, D. Trautmann, S. Sadovsky, and Y. Kharlov. Dense laser-driven electron sheets as relativistic mirrors for coherent production of brilliant X-ray and γ-ray beams. *Physics Report* 364, 359, 2002.

6. J. Eichler and Th. Stöhlker. Radiative electron capture in relativistic ion-atom collisions and the photoelectric effect in hydrogen-like high-Z systems. *Physics Report* 439, 1, 2007.

7. Kai Hencken. Transverse momentum distribution of vector mesons produced in ultra-peripheral relativistic heavy ion collisions. *Physical Review Letters* 96, 101–105, 2006.

8. L. Langer, S. V. Poltavtsev, I. A. Yugova, M. Salewski, D. R. Yakovlev, G. Karczewski, T. Wojtowicz, I. A. Akimov, and M. Bayer. Access to long-term optical memories using photon echoes retrieved from semiconductor spins. *Nature Photonics* 8, 851–857, 2014.

9. Leijing Yang, Sheng Wang, Qingsheng Zeng, Zhiyong Zhang, Tian Pei, Yan Li, and Lian-Mao Peng. Efficient photovoltage multiplication in carbon nanotubes. *Nature Photonics* 5, 672–676, 2011.

10. Li Qiong, D. Z. Xu, C. Y. Cai, and C. P. Sun. Recoil effects of a motional scatterer on single-photon scattering in one dimension. *Scientific Reports* 3, 3144, 2013.

11. Ping-Yuan Lo, Heng-Na Xiong, and Wei-Min Zhang. Breakdown of Bose-Einstein Distribution in Photonic, Crystals. *Scientific Reports* 5, 9423–9427, 2015.

12. B. Najjari, A. B. Voitkiv, A. Artemyev, and A. Surzhykov. Simultaneous electron capture and bound-free pair production in relativistic collisions of heavy nuclei with atoms. *Physical Review A* 80, 012701, 2009.

13. N. Artemyev, U. D. Jentschura, V. G. Serbo, and A. Surzhykov. Strong electromagnetic field EFFECTS in ultra-relativistic heavy-ion collisions. *European Physical Journal C* 72, 1935, 2012.

14. N. D. Benavides and P. L. Chapman. Modeling the effect of voltage ripple on the power output of photovoltaic modules. *IEEE Transactions on Industrial Electronics* 55(7), 2638–2643, 2008.

15. Peter Arnold. Photon emission from ultrarelativistic plasmas. *Journal of High Energy Physics* 2001, 057, 2001.

16. S. A. Klein. Calculation of flat-plate collector loss coefficients. *Solar Energy* 17, 79–80, 1975.

17. M. S. Tame, K. R. McEnery, Ş. K. Özdemir, J. Lee, S. A. Maier, and M. S. Kim. Quantum plasmonics. *Nature Physics* 9, 329–340, 2013.

18. M. W. Y. Tu and W. M. Zhang. Non-Markovian decoherence theory for a double- dot charge qubit. *Physical Review B* 78, 235311, 2008.

19. T. Pregnolato, E. H. Lee, J. D. Song, S. Stobbe, and P. Lodahl. Single-photon non-linear optics with a quantum dot in a waveguide. *Nature Communications* 6, 8655, 2015.

20. U. Becker, N. Grün, and W. Scheid. K-shell ionisation in relativistic heavy-ion collisions. *Journal of Physics B: Atomic and Molecular Physics* 20, 2075, 1987.

21. Y. F. Xiao et al. Asymmetric Fano resonance analysis in indirectly coupled microresonators. *Physical Review A* 82, 065804, 2010.

22. W. De Soto, S. A. Klein, and W. A. Beckman. Improvement and validation of a model for photovoltaic array performance. *Solar Energy* 80(1), 78–88, 2006.

23. Wei-Bin Yan and Heng Fan. Single-photon quantum router with multiple output ports. *Scientific Reports* 4, 4820, 2014.

24. W. M. Zhang, P. Y. Lo, H. N. Xiong, M. W. Y. Tu, & Nori, F. General non-Markovian dynamics of open quantum systems. *Physical Review Letters* 109, 170402, 2012.

25. Yu Zhu, Xiaoyong Hu, Hong Yang, and Qihuang Gong. On-chip plasmon-induced transparency based on plasmonic coupled nanocavities. *Scientific Reports* 4, 3752, 2014.

26. O. Khaselev. A monolithic photovoltaic-photoelectrochemical device for hydrogen production via water splitting. *Science*, April 17, 1998.

27. E. Kamal, A. Aitouche, R. Ghorbani, and M. Bayart. Fault Tolerant Control of Wind Energy System subject to actuator faults and time varying parameters, *2012 20th Mediterranean Conference on Control & Automation (MED)*, 2012.
28. Md. Faruque Hossain. Theory of global cooling. *Energy, Sustainability and Society* 6, 24.
29. Md. Faruque Hossain. Solar energy integration into advanced building design for meeting energy demand and environment problem. *International Journal of Energy Research*, 40, 1293–1300, 2016.
30. Yuwen Wang, Yongyou Zhang, Qingyun Zhang, Bingsuo Zou, and Udo Schwingenschlogl. Dynamics of single photon transport in a one-dimensional waveguide two-point coupled with a Jaynes-Cummings system. *Scientific Reports* 6, 8223–8225, 2016.
31. Qiong Li, D. Z. Xu, C. Y. Cai, and C. P. Sun. Recoil effects of a motional scatterer on single-photon scattering in one dimension. *Scientific Reports*, 3, 3144, 2013.
32. Masujima. Calculus of variations: Applications. *Applied Mathematical Methods in Theoretical Physics*, Wiley, Hoboken, NJ, August 19, 2009.
33. Reed M. Maxwell and Laura E. Condon. Connections between groundwater flow and transpiration partitioning. *Science* 353(6297), 377–380.
34. Scott Jasechko, Zachary D. Sharp, and Peter J. Fawcett. Terrestrial water fluxes dominated by transpiration. *Nature* 496, 347–350, 2013.
35. Jaivime Evaristo, Scott Jasechko, and Jeffrey J. McDonnell. Global separation of plant transpiration from groundwater and streamflow. *Nature* 525, 91–94, 2015.
36. Josette Masle, Scott R. Gilmore, and Graham D. Farquhar. The ERECTA gene regulates plant transpiration efficiency in *Arabidopsis*. *Nature* 436, 866–870, 2005.
37. Tobias D. Wheeler and Abraham D. Stroock. The transpiration of water at negative pressures in a synthetic tree. *Nature* 455, 208–212, 2008.

9 Self-Sustaining Urbanization and Self-Sufficient Cities in the Era of Climate Change

Negin Minaei

CONTENTS

ABBREVIATIONS

AHP	analytical hierarchy process
CGBC	Canada green building council
DRR	disaster risk reduction
EPA	Environmental Protection Agency
GEF	global environment facility
GPSC	Global Platform for Sustainable Cities
LEED	Leadership in Energy and Environmental Design
NMT	nonmotorized transport
PPP	public participatory planning

PROMETHEE	preference ranking organization method of enrichment evaluation
SDGs	Sustainable Development Goals
SMM	sustainable material management
TOD	transit-oriented development
UHI	urban heat islands
UNDP	United Nations Development Programme
UNEP	UN Environment Programme
ZEB	zero-energy building

9.1 INTRODUCTION

Urbanization is a complex subject that cannot be discussed in all its dimensions and aspects in a chapter or even a book. In this chapter, the focus is on self-sustaining urbanization. It is possible to mention only a few aspects with higher priorities that require immediate attention and long-term investments by cities. Figure 9.1 illustrates how cities can progress from resilient to sustainable, smart and eventually self-sufficient cities using a system thinking approach.

United Nations (UN) projected by 2050 at least 70% of the global population will live in urban areas, or, two-thirds of humanity, which equals 6.5 billion people. This will cause more strains on cities and their urban infrastructure as well as the services for which they were planned, designed, and built. Although population growth was identified as one of the contributing factors to climate change (UNHABITAT,

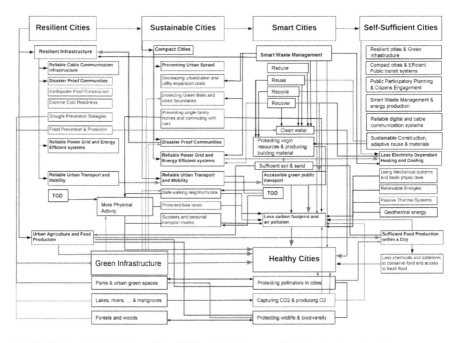

FIGURE 9.1 Self-Sufficient City. (Minaei, N., Drawn with Google Drawings.)

2016b, p. 2), it seems reversing the migration pattern, from rural areas to urban areas is more complicated than finding solutions for climate change because cities are still the ones that generate wealth, employment, and human progress. These solutions are to lead and equip cities with the type of infrastructure that enables them to become resilient, resource-efficient, sustainable, well-managed, and ultimately self-sufficient. Surely, significant redistribution of wealth and political arrangements to suit them are part of the required reforms. Urbanization has a direct impact on global environments and that is the reason that "Sustainable Urbanization" was identified as a priority (GEF, 2018). Many international organizations have concentrated on the urbanization process in recent years; organizations such as United Nations (UN) and its associated agencies including UNDP, UNHABITAT, UNEP, UNIDO, World Economic Forum, World Bank, European Bank, and newly established organizations such as C40 CITIES, 100 Resilient Cities, Global Platform for Sustainable Cities (GPSC), Smart Cities Council, and the list goes on.

This chapter introduces the concept of self-sufficient cities and self-sustaining urbanization. Because these concepts are somewhat new and this book is written for engineers, I limit the terminology and philosophy of these new planning concepts to the everyday language and start with simple dictionary definitions. We explore the similar terms that have been used in this context and explain the fundamental matters such as Sustainable Development Goals (SDGs) and resilience in cities to better define the self-sufficiency. In the second section, possible fast-track solutions that have been proven to be efficient in achieving sustainability by successful cities across the globe will be briefly reviewed to provide a simple understanding of the most important factors for non-planning professionals.

9.1.1 TERMINOLOGY AND DEFINITIONS

9.1.1.1 Self-Sufficient, Self-Reliance, or Self-Sustained

Oxford Dictionary defines self-sufficient as "*Needing no outside help in satisfying one's basic needs, especially with regard to the production of food.*" (2019), while Cambridge Dictionary says "*the ability to provide what is necessary without an outside help,*" which can include economy or a person; The second definition is a more comprehensive definition (Cambridge Dictionary, 2019). In this chapter, "Self-Sufficient Urbanization" exceeds urban food production and looks at a city as a complex and comprehensive system with a main scope, self-sufficiency. A system that needs to be self-reliance, self-sufficient, and self-sustained with the ability to survive in harsh conditions such as extreme weathers and to respond to its inhabitants' needs without receiving immediate help from external sources; it is self-dependent in all aspects from skill sets and human resources to natural resources and food production. This brings us to the concept of "Resilience." UNHABITAT (2016a) has defined resilience an ability of human settlements to "*withstand and recover quickly from any plausible hazards.*" Recovering to predisaster standards is often costly and time-consuming; sometimes it takes decades. In urban context, a system comprised of communities, individuals, businesses, institutions, and other systems within a city. A self-sufficient system needs to foresee possible future events–both technological advancements, and the transformation they cause–to the consequences of climate change; This system has been planned and prepared to face any possible crisis, shock, stress, or hazards; in other words, it needs to be resilient first. 100 Resilient Cities (2019a) defines urban resilience the ability "*to survive, adapt, grow regardless of acute shocks and chronic stress.*" Therefore, being resilient is a prerequisite for becoming sustainable or eventually self-sufficient.

Technology alone cannot change the way cities work because it is expected in Smart Cities to overcome most of the problems cities currently have and achieve high levels of liveability and sustainability for urban inhabitants as argued comprehensively by Minaei (2017 in Huston, 2019; Hamnett in Moore, 2014). A self-sufficient system needs a shared vision, well-thought plans with clear strategies, achievable targets, and feasible scenarios that are clearly communicated to citizens in advance or in the best-case scenarios are the results of public participatory planning (PPP) or community planning as conscious citizens are the pillar of a Smart City and society. It is the social and community communication and capacities for cooperation among governments, stakeholders, and citizens that can strengthen the civic engagement, the core of a Smart City.

There is little literature about self-sufficient urbanization. Barcelona's chief architect, Vicente Guallart was perhaps the first person who used the term and envisioned Barcelona as a self-sufficient city in the context of Smart City with environmental urban solutions (March and Ribera-Fumaz, 2016). In a lecture series offered by the Department of Urban Studies and Planning at the MIT ("Sustaining Cities" as part of the "MIT World Series of Changing Cities"), Judith Layzer addressed decreasing demands in cities could be the key factor to achieve strong sustainability; Corburn identified having an environmental justice framework to look at cities from a health point of view (social side) was more important than ecological sustainability; and Zegras explained why the access to opportunities by providing equitable public transport was the key to achieve sustainable development (Layzer et al., 2013). These contributions are only small parts of what is required for a truly sustainable city.

9.1.1.2   Self-Sustaining Urbanization and Self-Sufficient Cities

Different city concepts and multitude initiatives have looked at the sustainability of urban developments during the past decades. Most of them aim to compete and increase their status in the global cities' hierarchy by achieving sustainability and advancing and upgrading economic, environmental, and social conditions although their principles and frameworks slightly differ (Minaei, 2017a; De Jong et al., 2015). Among them, "Resilient Cities," "Digital Cities," "Smart Cities," and "Sustainable Cities" are the chief concepts. Other concepts view cities with a particular lens such as environment for "Eco Cities," "Low-Carbon Cities," and "Green Cities"; technology for "Information Cities," "Digitized Cities," (Landry, 2016) or "Intelligent Cities"; education for "Learning Cities" (Kearns, 2012), "Knowledge Cities," "Creative Cities," (Landry, 2012) or "Innovation Cities" (2ThinkNow, 2018); health and well-being for "Healthy Cities" (WHO, 2009) and "Liveable Cities" or a recent concept of "Data Cities" (Lund Humphries in Jackson, 2019). By far, the "Sustainable City" concept is the oldest concept being used (since 1996) as the most common category

(546 published articles as opposed to the "Smart City" with 222 articles), which stands at the top of all other concepts (De Jong et al., 2015). At the core, all aim to achieve sustainable development, but the degree to which they concentrate on social and environmental aspects differs. For example, principles and values of a healthy city are: equity, participation and empowerment, working in partnership, solidarity and friendship, and sustainable development (WHO, 2009). The definition of a Healthy City can be similar to the "Learning City" because both concepts are shaped around community resources, ensuring economic development through partnership, and providing access to opportunities and skill development and overall continuously improving the physical and social environment relying on community and the existing resources (Kearns, 2012, p. 376). Cities do not need trendy concepts and super advanced technologies to gain sustainability; they need to invest on the natural systems, people, and societies to enable them to move toward self-sustaining urbanization. That collaboration, empowering citizen engagement, conserving the existing resources, and expanding them to the maximum potential within a resilient infrastructure seem to count good examples of self-sufficiency in cities.

9.1.1.3   Sustainable Urban Development

The increasing population growth in urban areas and unequal distribution of population and density in cities resulted to some areas with higher density and lower quality of housing, which make those areas vulnerable to natural hazards and climate change. This is particularly the case in mega-cities with the population overload (Malalgoda et al., 2013). Much degradation of the natural ecosystem is caused by human activities, and it is the sustainable urban development that works with all three necessary dimensions of economy, environment, and society (Yigitcanlar et al., 2019). Now, most citizens are aware of climate change and the possible consequences it can bring to cities. Devastating and unpredictable natural hazards are happening more frequently and more severely. Although scientists can predict such extreme events may happen, the truth is no one can foretell the location of the next disaster. In some cases, city officials were aware of the catastrophic events in advance, but they could not warn and prepare residents because of the lack of efficient mechanisms and connectedness to communities (Lejano, 2019). The importance of "emergency preparedness" for communities and equipping a city's resilient infrastructures to prepare for future hazard (e.g., earthquakes, floods, droughts, forest fires, water scarcity, air pollution, etc.) is undeniable. Having resilient urban environments is in fact the first and most important step to be taken toward achieving a sustainable city. During the past decades, urbanites have been dependent on the import of goods and services with high carbon footprints, particularly fruits and vegetables transportation that account for 50% of the total carbon emissions (Weber and Matthews, 2008 in Wakeland et al., 2012, p. 212). Most of the food that is available in cities either comes from rural areas or is imported from other countries. Cities no longer produce the food they need, which could be the result of insufficient agricultural lands (Churkina, 2016). Sustainable food production in cities (urban agriculture or urban farming) has been one of the important actions cities started to take in recent years, and there is a growing body of literature on methods of reducing food-related greenhouse gas (GHG) emissions such as changes in consumptions patterns, dietary choices, distribution systems, and food production (Galli et al., 2017, p. 384). Repurposing derelict lands in urban areas has benefitted many environmentalists and nongovernmental organizations (NGOs). In some cities, municipalities have assigned parts of urban parks or neighborhoods' lots to urban farming or permitted towers and high-rise buildings to use rooftops and balconies to grow food. Scarcity of water and energy are also among the main challenges of all cities, finding the optimum solution depends on many factors, including geography, climate, environmental characteristics, and within the financial parameters of a city (Gardner, 2016 in Alkhalidi et al., 2018). In designing new residential complexes, architects and landscape architects look at low-tech solutions to decrease energy dependence. For example, in the on-site farms of a recent Dubai sustainable city project, simple fans were employed without any air-conditioning systems (Fullychargedshow, 2017).

9.1.1.3.1   Sustainable Development and Its Complex Assessments

The first step to move toward resilience is to learn about the current state of a city by benchmarking and proper assessments. Assessing sustainability in cities has been conducted by applying different tools or comprehensive models in different scales: from building (super-micro), parcel (micro), neighborhood/suburb (mezzo), city/region (macro), (supra)nation, and a (super-macro) scale (Fredrick, 2014 in Yigitcanlar et al., 2015). This means although UN has provided targets and indicators, these inconsistencies between measures used by different cities makes the annual reports submitted to UN difficult to compare. Unsuccessful attempts to achieve the eight voluntary millennium goals by different countries by 2015 led the UN to introduce the seventeen SDGs, which are obligatory. Countries that signed the Paris Agreement must achieve these goals by 2030 or present evidence to prove they had solid plans for; they must annually submit some progress reports to the UN. The UN's SDGs cover the three dimensions of sustainability including economic, environmental, and social; Goal 11 is assigned to "Sustainable Cities and Communities" but the fact is the built environment is the context for the all other sixteen goals, too. In other words, "No Poverty," "Zero Hunger," "Quality Education," "Climate Action," "Affordable and Clean Energy," and other goals cannot be separated from cities as cities are their context and albeit SDGs are all intertwined too.

Most cities are not able to measure their sustainability level, and they rely on private companies to provide those measures and produce reports for them, which questions the credibility of those reports. The most common reasons for not being able to measure sustainability index in a city is first, lack of required and credible data (most cities have not developed mechanisms to collect city's data, sort, and analyze them and do not have Smart City's infrastructure) and second, lack of skill sets to identify the proper assessment model by city's staff members to conduct the assessment.

The fact that various assessment models with different indicators are available does not decrease the complexity of selecting the right one. For example, Gervásio and da Silva (2012) proposed a probabilistic decision-making approach using *prefer-ence ranking organization method* of *enrichment evaluation* (PROMETHEE) and analytical hierarchy process (AHP) to help governments assess their infrastructure on a more reliable basis. Carli et al. (2018) used AHP to identify a multicriteria decision-making assessment that best fits for metropolitan cities based on the thirty-five indicators of the Sustainable Development of Energy, Water and Environment System Index (SDEWES) framework. Kilkiş (2016) provided exact measures and formulas to calculate each index of the SDEWES for a city.

Nevertheless, that does not mean that cities in different countries are moving toward a sustainable future! GPSC started working with twenty-eight cities across eleven countries to support them improve their sustainability focusing on the *sustainability indicators and tools, integrated urban planning and management, and municipal finance* (GPSC, 2018). Many noncapital cities in the Global South are not even aware of what their obligations are and have not started taking any actions.

9.2 FAST-TRACK EFFICIENT APPROACHES TO ACHIEVE SELF-SUFFICIENCY IN CITIES

CASE STUDY: TORONTO

Pioneer cities such as most smart and resilient cities have developed plans to decrease their environmental impacts and, therefore, slow down the process of climate change and simultaneously prepare for the climate change by adaptation and mitigation. A good example is Toronto. Toronto has a climate action strategy that was approved by the City Council in 2017, and it aims to reduce greenhouse gas (GHG) emissions: 30% by 2020, about 65% by 2030, and ultimately 80% by 2050 (City of Toronto, 2019a). That means considerable changes should be introduced in the ways construction, transportation, and energy industries and waste management systems work. The mayor of Toronto, John Tory joined the C40Cities and committed to decrease Toronto's GHG emissions by taking actions toward preparing it for the climate risks, prioritizing the shocks and stresses based on the severity of them, and monitoring the carbon dioxide emissions by source (C40CITIES, 2019).

Toronto is one of the three Canadian cities that have taken the action to join the global network of "100 Resilient Cities" and was assigned a chief resilience officer to lead its resilience strategy against the main shocks of rainfall flooding, blizzards, and heat waves and main stresses of economic inequality, energy insecurity and power outage, and lack of affordable housing (100 Resilient Cities, 2019a, 2019b). The Chief Resilient Officer established a ResilientTO section to prepare people, communities, and businesses in Toronto to be resilient by development and implementation of the first Toronto Resilience Strategy (City of Toronto, 2019b). Involving citizens and asking them to share their personal stories of resilience, mapping the type of stresses and shocks they have experienced across Toronto, and completing surveys to get more information

(Continued)

and engage inhabitants of Toronto are the steps that ResilientTO program have taken to ensure citizens are informed. Unfortunately, Rockefeller center, who funded the 100 Resilient Cities, announced that he would no longer fund this initiative as they could not achieve their goals by their timeline (Pitt, 2019). An unofficial source from the City of Toronto mentioned the city has a plan to continue the role.

Toronto commenced upgrading its green infrastructure by adaptive reuse of old infrastructure (the Beltline Trail and West Toronto Rail path) and introducing an initiative called "Tree For Me," which provides free native plants to citizens to plant them in their back yards to increase its green canopy and decrease its carbon footprint (Toronto Park and Trees Foundation, 2019).

Live Green Toronto under the Department of Water and Environment of the City of Toronto has a variety of initiatives and incentives to promote sustainable development, which span from the Home Energy Load Program (HELP) to Smart Commute, Pollinator protection, Eco-roof, green local businesses, and many more. This department has trained more than 14,000 volunteers that actively inform and educate the public on waste management, water management, and everything to do with the built environment in the city of Toronto.

9.2.1 COMPACT CITIES AND LAND-USE CONFIGURATION

"Compact Cities" prevent urban sprawl by employing high-density developments; high-density development is not the same as vertical growth (Foster, 2015). Compact Cities are an optimal urban form for a sustainable city and can reduce energy wastes, conserve energy, and reduce the expansion costs as building new infrastructure (roads, pipes, electric grid, sewers, broadband, etc.) and expanding horizontally is quite costly. Additionally, studies showed that the amount of carbon that is stored in one unit of urban area depends on its form whether it is compact or sprawled (Churkina, 2016). To prevent urban sprawl means to minimize the prevalence of inequity and single-family homes in the suburbs and the use of cars to commute to a city center. In British sustainable cities such as Bristol, there are clear city boundaries, which are often protected by "green belts" and "park and ride" stations right on the edge of the city; people have to park their cars in those stations and commute to a city by public transport modes.

Applying a mixed land-use approach in a city's masterplan has been a remedy to bring comfort to citizens. In addition, size of the street blocks has a direct impact on peoples' choices of transport mode. Bigger block size discourages people to walk, while denser and smaller blocks bring people to the streets to walk or ride a bike; small European towns such as Venice, Lisbon, Florence, and Prague are good examples of walkable cities because people can walk everywhere and have access to all amenities without needing to use their cars. In most "organic forms of growth" in cities, commercial and service uses were assigned to the street level and residential and office uses were placed on top of these spaces. In modern and new cities with "geometrical urban patterns" or "grid systems," often big blocks and lack of mixed-use properties force people to use their cars to travel to other parts of a city, such as a plaza the case for most US and Canadian cities, only to buy their basic groceries, which is clearly a nonsustainable approach in urban design.

9.2.2 Transit-Oriented Development (TOD) and Walkable Neighborhoods

Urban studies have shown when the density increases, people tend to use public transport or nonmotorized transport (NMT) almost by 20% (World Bank Sustainability Blog, n.d.). That has been a good sign to develop an approach called TOD, which is about brining compact and mixed-use developments around public transit facilities. This approach was first applied in China by the World Bank to create "Livable Cities" and has found some followers like Jennifer Keesmaat who suggested to bring affordable housing on and around metro stations across Toronto. One clear benefit of TOD is the opportunity it provides to design walkable neighborhoods and car-free areas, which leads to accessible neighborhoods and ultimately less energy consumption and less air and noise pollution. Many studies such as Heath et al. (2006) have shown the impacts of land use and urban design on physical activities and health. Walkable neighborhood designs and access to basic amenities such as grocery stores, library, schools, and health clinics could increase walking and, therefore, decrease obesity in young children and adults. Also transport studies illustrate the most sustainable forms of mobility are walking and biking; this means planning and designing proper infrastructure for these modes, such as assigning bike paths and scooter lanes, is required so high-speed traffic and safe movements can naturally flow across the city and region. On a larger scale, urban scale, advancing high efficiency and clean public transport modes increases the NMT plus reduces noise and air pollution as well as incidents with vehicles in cities.

9.2.3 Sustainable Infrastructure

Patrick Geddes (1915) who was initially a biologist interested in town planning was the first person who looked at a city as a living organism containing human and non-humans and the built environment. He combined urban planning with ecology and invented regional planning, which later led to new fields such as ecological urbanism and landscape urbanism. The relationship between natural resources, including the land and its biodiversity as a context for cities to grow, and the land-use patterns with the population are still valid. The more naturally resourceful a city, the more population it accommodates (Parris et al., 2018). Unfortunately, urbanization and land-use changes have caused extinction among many species, which reversely impacts the stability of the ecosystem in the area and beyond and in the long term. To overcome these problems, Kristen Parris et al. (2018) defined seven ecological principles, or as they call it seven lamps of planning, to protect biodiversity in cities, which are: identifying the protection areas, connecting habitats to allow movements of different species, building ecological elements to provide habitats, sustaining eco-systems (energy cycling, nutrient, and water), and carefully thinking and designing urban form and urban infrastructure to avoid negative impacts on the wildlife popu-lation and biodiversity in cities while creating new ecosystems to allow new species develop their own habitats. The built environment base for these lamps are heritage, mobility, "heterophilia," integrated water management, neighborhood, safety and well-being of inhabitants, and finally, urban renewal. This brings us to the green infrastructure in pioneer sustainable cities.

9.2.3.1   Green Infrastructure and Urban Space Regeneration

Numerous studies have proved that green infrastructures, which are often natural and self-sustained, are the only way for cities to become resilient and to survive the impacts of climate change (McPhearson et al., 2015; Tillie and van der Heijden, 2016). Lakes, rivers, mangroves, woods and urban forests, parks and urban green spaces, and the newly emerged phenomena of "freeway caps" or "freeway cap parks" are the most common forms of green infrastructure. They act as lungs for cities to breathe in all sorts of emissions and breath out oxygen, which is essentially what all living creatures need. They are also effective in decreasing the urban heat islands (UHI) effect and keeping cities cool, which means conserving energy. Repurposing old infrastructure such as old rail lines, car parks, or closed freeways and adding a green space to it created projects like rail-to-trail or freeway-to-Boulevard. Recently adding a green deck in the air space right above a freeway has become a new trend among megacities in Europe and the United States (e.g., Promenade Plantee in Paris, Rose Kennedy Greenway in Boston, and High Line in New York) (Houston and Zuñiga, 2019). These freeway cap parks can reduce the pollution (air and airborne, storm water, noise, etc.), create a micro-climate, mask the high-speed traffic below, provide habitats for animals, and a green corridor to help biodiversity of the area plus reintegrate communities that were originally divided by that freeway.

Ensuring a city's environment and its urban ecosystem is protected and well-conserved can be achieved by different initiatives, most notably, incorporating green industries and decarbonizing urbanization, low-carbon technologies to gen-erate power and to operate a city's transport system, maintaining and protecting biodiversity, smart solid waste management, and planning for a sustainable urban future (GPSC, 2018). Certainly, green infrastructure solutions differ from region to region based on their climate, typography, vegetation, and other environmental characteristics.

9.2.3.2   Sustainable Energy: The Transition

Decarbonizing the power generation and transition from fossil fuels to renewables, which decreases the total amount of power generated, have been one of the main initiatives that governments have been working on. De-carbonization of electric grid has been the force for the change of heating systems to electric based ones in build-ings in recent years (O'Dwyer et al., 2019), and although that may seem sustainable, it is not self-sustaining and reliable. With the push toward electric vehicles, which increases the pressure on the electric grid, a more efficient and intelligent energy man-agement system is required to generate power from multiple and different sources.

The European Union (EU) has provided an economic and political base to invest on clean energy and efficient energy consumption but that is not easily achievable. For instance, despite changing political control in London (UK), clean energy targets (25% by 2025 and 50% by 2050) as part of the "Green Growth" has stayed the same; they must be provided from local low-carbon sources (Webb et al., 2016). The EU's "zero-energy buildings" (ZEB) concept envisioned all new buildings should be nearly zero energy by December 2020; although this has not happened by 2019, a new concept of "nearly zero-energy building (NZEBs)" has been commonly used instead to decrease the gap and make the "European Smart Cities and Communities Initiative" achievable (O'Dwyer et al., 2019). Studies have shown heating and cooling buildings is one of the three main reasons for excessive energy consumption. Globally, about 40% of the energy is consumed by buildings (De Rubeis et al., 2018). In Canada, in 2016, in the residential section, most of the consumed energy belonged to space heating (62%), space cooling (2%), water heating (20%), and lighting (4%) (Natural Resources Canada, 2016, p. 99). That decreased only by 1% in space cooling and 1% in water heating in 2017 (Natural Resources Canada, 2017, p. 117), and the rest remained the same. Overall 86% of the energy consumed for heating could be conserved to some degree depending on the season and the proper design of buildings. In some cases, behavioral patterns rule the energy consumption. A tangible instance are the rental apartments in older buildings in Ontario, which are almost always too warm, and because residents cannot decrease the indoor temperature, they open their windows to cool them down, which wastes lots of energy. Having residents from different parts of the world does not help too because their comfort temperatures are different but due to the old heating systems all must adapt with the same often high thermal comfort temperature. While the standard comfort temperature globally is between 18°C and 23°C, in the United Kingdom, the minimum starts at 13°C (HSE, 2018). I vividly recall that the heating system in our offices were off till the indoor temperature dropped under 13°C; some international colleagues brought a hidden electric heater to warm up their feet and hands to be able to work. The same was the case for residential spaces. particularly in the private sector where most landlords would not turn on the heating often because fuel was expensive (they thought they saved on utility bills and conserved energy but were not aware their tenants found other solutions to keep them warm such as hidden electric heaters or thermal blankets, etc.).

Recent innovations in the field of sustainable energy generation in remote areas, such as Africa, which do not have access to electric grids, seem to be very efficient for the rest of the world as well. These innovations mainly rely on mechanical operations and the laws of physics to generate power and often do not have negative environmental impacts. For instance, small portable wind turbines or the turbine that can be placed in a stream to generate electricity or the gravity light, which provides light by pulling a string to bring a weight up. The gravity light idea was later bought from the African inventor, adopted by a British company, and many products, including a USB system to charge phones and bring light using the same technology, have been developed. A new material called "snow-based triboelectric nanogenerator, or snow TENG" was recently tested by scientists at the University of California, Los Angeles, which could generate electricity from snowfall and could coat solar panels to generate electricity, even in winter.

9.2.3.3 Urban Water Management

Perhaps sustainable urban water management is the most important infrastructure that should be carefully studied, designed, and implemented for cities, especially when we remind ourselves only 3% of the world's water is drinkable. It includes the freshwater supply, rainwater collection, environmental protection, proper sewer systems, catchments, and landscape design to prevent the risk of flooding in cities. Mangroves and lakes (both natural and artificial) can retain excess surface water in case of severe rains and are considered green infrastructure. According to UN Environment (2017), these "blue forests" are declining fast, because they have a large carbon sink capacity. Unfortunately, mangroves as a forgotten green infrastructure are diminishing in most cities while they could play such important roles for the survival of cities by capturing carbon dioxide and GHG emission as well as flood prevention, filtering runoffs, and providing micro ecosystems as habitats in urban areas. Singapore and Berlin as internationally known cities for their water management systems have used green infrastructure on a large scale to collect reclaimed and storm water, treat it and reuse it for water supply. Tianjin eco-city and Melbourne have facilitated irrigation of public green spaces with nonconventional water, so they could conserve potable water for people's consumption (Liu and Jensen, 2018). It is strange and perhaps unwise that humans do not pay attention to the natural solutions that have proved efficient for centuries and instead try dangerous solutions such as "geoengineering" with known negative and long-term environmental impacts on nature and wildlife just to respond to a short-term problem (Dunne, 2018).

9.2.4 Sustainable Architectural Design and Construction

Construction process starts from the site selection to the interior design of a space. A sustainable site has had former developments and has been already equipped with proper access to utilities, including existing power grids, storm water run-off (Kurle, 2017), green urban spaces, and public urban facilities such as schools, public transport, libraries, and clinics. Planning for mechanisms to offset the carbon on site should be considered even before the design phase. This has become a recent trend for most new complexes by generating energy on site, recycling wastes on site, and ensuring the landscape design has solutions for rainwater capture, sustainable irrigation and, in some cases, producing food on site. Surely, alternating land-use from agricultural land to any other use is not sustainable, considering the global lack of agricultural soil. If the selected site needs demolition, one sustainable practice is to carefully upcycle the existing construction and demolition (C&D) materials from glasses and window frames to the metal, bricks, concrete, and even trees. Lack of original resources such as sand and soil for concrete production and the concept of "emergy" —the memory of energy formerly was used in a material (Odum, 1988)— were the starting point of reusing and recycling construction materials. The US Environmental Protection Agency (EPA, 2017) promotes sustainable material management (SMM) by informing the public on the process (reduce, reuse, recycle, and rebuy C&D materials) and providing enough information and guidelines on the types and the benefits of materials. Figure 9.2 illustrates the way with which a linear system of consumption can transform to a nonlinear sustainable system (Figure 9.2).

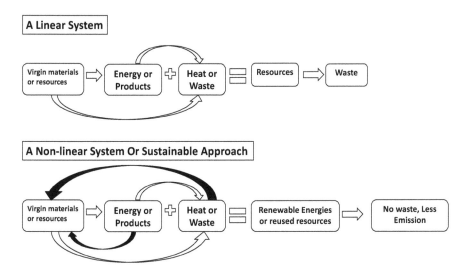

FIGURE 9.2 Linear and nonlinear systems and the sustainable approach. (From Minaei, N., *Implications of Energy Use and How to Reduce the Impact*, University Lecture, Royal Agricultural University, Cirencester, UK, 2013.)

Thinking and planning for energy efficiency and water efficiency in buildings are good practices that should be encouraged. In most recent sustainable designs, using active energy systems such as solar cells and solar thermal panels as well as geothermal energy have been prevailed. Most sites allocate an area to collect rainwater in specific facilities and treat their grey and black water. In a new smart project in Dubai, grey and black water stream down the site through biofilters such as papyrus plants reaching the treatment facility and later used for irrigation of crops that are gown on site (Fullychargedshow, 2017). In addition to the natural filtering, the current water can cool the surrounding environment; all can be achieved by sustainable landscape design. Recycling and upcycling construction material to prevent generating construction wastes were set up on this site as well to provide the opportunity of recycling materials after any renovation or refurbishment.

Traditionally in historical architecture of the Far East (China) and the Middle East (Iran), it was common to use the ground and build underground so levels of a building could stay cold in summer and warm in winter only working with the natural ventilation and air flow patterns (Soflaei et al., 2017). The traditional construction has progressed as well; the straw-bale buildings (United Kingdom) and the "natural building" movement (United States), which rely on passive wall systems, natural current air, natural light and sunshine, and capacities of natural materials seems to create energy-efficient buildings.

The most well-known form of sustainable construction is the eco-tech architecture and green architecture in which climatic factors and typography of the site directly influence the design of a building and its energy performance. Such architectural styles have led to emergence of green buildings, Leadership in Energy and Environmental Design (LEED) and the zero-carbon building standards as was briefly discussed before. ZEBs have almost seventy different standards and definitions across the world, which means as a category they need a universal definition (De Rubeis et al., 2018) if measuring the energy efficiency or a comparison of cities' energy consumption based on their carbon footprint should be reported. LEED Canada have addressed these topics: "*sustainable sites, water efficiency, energy and atmosphere, material and resources, indoor environmental quality, innovation in design and regional priority*" for new constructions and major renovations after 2010 (LEED, 2010, p. xiv). New buildings can earn the "Zero Carbon Building-Design certification" by illustrating the carbon balance, the ventilation system, and the methods they applied to ensure their building envelop was highly efficient plus the way they incorporated an on-site renewable energy system (CGBC, 2017).

Using active and passive energy systems in buildings to decrease the energy consumption for heating and cooling and to conserve energy have been the main approach toward a more sustainable construction. Passive heating harnesses solar radiation to gain internal heat and stores it in different elements such as solar mass, solar wall, Trombe wall, water wall, roof garden, roof pond, green house, patio, etc. It can, to some degree, provide heating and cooling. Architectural design considerations such as improving insulation, adding filtering spaces, covering facades with vegetation, and upgrading utilities to energy-efficient products and systems, including heating and cooling with hydronic systems can benefit buildings to a large degree (Minaei, 2017b). A passive design approach is an architectural design style that can improve thermal comfort, decrease energy consumption in buildings and energy costs by 85%, and reduce carbon dioxide emissions by 65%. It is subdivided to three main methods: "direct gain," "indirect gain," and "isolated gain." The city of Vancouver has officially recognized passive design as smart solution to build smart high-performance buildings (City of Vancouver, 2009). It is crucial to note that because climatic factors differ from location to location, not all solutions that are employed in one climate are necessarily applicable in another one. For example, design options that are useful in a hot and dry climate such as white roofs to cool down buildings and prevent UHI are not necessarily good for a cold and wet climate like Toronto or south-facing windows, which can warm and heat the space during the day and decrease the energy consumption at night, are not desirable for example in Dubai because avoiding sun is the best option for there. This has been the common method: Architects have been trained to find the optimum form and function by design. A recent modern example of sustainable construction is the design of a complex of 35,000 m² with the UN seventeen SDGs in mind, called the UN 17 Eco-village in Copenhagen, which is designed to accommodate 830 people and provide a hundred unskilled jobs. It grows most crops on site both in greenhouses and on roofs (UN Environment, 2019).

To truly be self-sufficient, cities should not overly rely on one source of energy and power such as electricity; but sadly, electrification of cities has been the latest trend in both transport and building industry. A power outage for couple of days in a cold season is enough to show the degree by which our cities are fragile and are not self-sufficient. It is a matter of survival first and then pursuing fancy technology and city brands like Intelligent City or Smart City. When people cannot heat their homes, cook, have access to water, or even use their electric vehicles to leave their places to go to a safer ones after a simple power outage, living in an electric Smart

City sounds meaningless regardless of the international reputation and high ranks in the global hierarchy. This is the type of situation that cities should plan for and be prepared, but even Smart Cities like Toronto are not ready for such small incidents. During winter 2018, Red Cross provided a warm room in a tower in Toronto, which lost power due to the flooding of the basement that shut down the whole building, and hundreds of people did not have access to basic human needs from showers, drinking water, and warm food to heating, electricity, or access to phone.

When it comes to self-sufficient homes, most architects employ climatic factors, orientation, and all principles in eco-friendly architecture: using natural material, designing an off-grid house with minimum environmental impacts, and relying on both passive and active energy systems such as passive heating, installing solar panels, backing up the heating system with either biodiesel or biogas fuels as well as a mechanical water supply system (Dezeen, 2019). A recent modeling and energy analysis of a ZEB in Italy, using EnergyPlus, DesignBuilder and thermographic images, illustrates the range of technologies a ZEB should have to be energy self-sufficient: a biomass boiler, a solar thermal plant, a photovoltaic system, a clean water tank, a domestic hot water tank, a rainwater reuse system, a black water reuse system, and a well for water supply (De Rubeis et al., 2018).

9.2.5 SOCIAL AND ECONOMICAL SUSTAINABILITY

It has been reported that low-income, elderly, and minorities are among the most vulnerable groups in urban population that are exposed to the devastating impacts of climate change. That means to achieve social equity and economic equality, cities must have plans to support these groups before, during, or after a natural hazard happens including extreme cold, flooding, storms, and more.

Local governments face some challenges that diminish their ability to create resilient and, therefore, sustainable cities (Malalgoda et al., 2013). To name just a few:

- Financial limitations to undertake activities to improve the resilience of their city;
- Lack of professional human resources and up-to-date knowledge about disaster risk reduction (DRR) and identification of their vulnerabilities.
- Lack of long-term strategies and plans to be followed due to staff change or policy change in higher levels of management.
- Most cities and municipalities have focused on the recovery stage rather than planning and preparation stages.
- Lack of collaboration among different departments of a municipality or the governmental body as well as insufficient communication NGOs to prevent parallel working and clearly define responsibilities.

9.3   CONCLUSION

Self-sustaining urbanization and self-sufficient cities are novel concepts that have not been discussed in the literature. This chapter uses a simple definition and builds on it with the facts and knowledge from scientific research to shape a concept for self-sustaining and self-sufficiency. Self-sustaining urbanization is about the mechanism that help an urban environment gain resilience and sustainability and ultimately self-sufficiency. The first step is to benchmark and assess the current state of a city by collecting reliable data, as has been recommended for smart cities, too, analyzing and identifying priorities to plan for investments and pursuing actions on weaknesses and threads while considering potentials and opportunities. As explained, resilience is the number-one priority for all cities. Because Resilient City and Sustainable City are known elaborated concepts, it is important to benefit from the knowledge and experience of professional organization such as the UN and other international NGOs that offer professional services to help cities progress toward sustainability. Surely, investing in some pillars of the Smart City such as digital infrastructure and conscious citizens can improve the situation and accelerate the process because without updated and reliable data, planning and implementing can take a long time. In the second section of this chapter, we reviewed the most efficient and successful approaches and best practices including land-use configuration, TOD, walkable neighborhoods, green infrastructure to manage water, energy water and construction industry with traditional architecture and passive design solutions. To be self-sustaining, perhaps the number-one solution is to use natural laws of physics and mechanical systems that do not rely on electricity or fossil fuels. Cities should not rely on online digital technologies or electrical utilities only. They should have manual mechanical backup systems that could be used in case of unforeseen disasters.

CONFLICTS OF INTEREST

In accordance with Taylor & Francis policy and my ethical obligation as a researcher, I am reporting that I do not have a financial and/or business interests in and have not received any funding that may be affected by the research reported in the enclosed chapter.

BIBLIOGRAPHY

2ThinkNow, 2018. Innovation Cities™ Index 2016–2017: Global. Available at: https://www.innovation-cities.com/innovation-cities-index-2018-global/13935/

100 Resilient Cities, 2019a. What Is Urban Resilience? Available at: http://100resilientcities.org/resources/#section-1

100 Resilient Cities, 2019b. Toronto's Resilience Challenge. Available at: http://www.100resilientcities.org/cities/toronto/

Alkhalidi, A., Qoaider, L., Khashman, A., Al-Alami, A.R. and Jiryes, S., 2018. Energy and water as indicators for sustainable city site selection and design in Jordan using smart grid. *Sustainable Cities and Society*, 37, pp. 125–132. doi:10.1016/j.scs.2017.10.037.

C40 CITIES, 2019. Toronto. Available at: https://www.c40.org/cities/toronto

Cambridge Dictionary, 2019. Meaning of "self-sufficient" in the English Dictionary https://dictionary.cambridge.org/dictionary/english/self-sufficient?q=Self-sufficient

Canada Green Building Council (CGBC), 2017. Zero Carbon Building Standard. Available at https://www.cagbc.org/zerocarbon

Carli, R., Dotoli, M. and Pellegrino, R., 2018. Multi-criteria decision-making for sustainable metropolitan cities assessment. *Journal of Environmental Management*, 226, pp. 46–61. doi:10.1016/j.jenvman.2018.07.075.

Churkina, G., 2016. The role of urbanization in the global carbon cycle. *Frontiers in Ecology and Evolution*, 3, p. 144. doi:10.3389/fevo.2015.00144.

City of Toronto, 2019a. TransformTO. Available at: https://www.toronto.ca/services-payments/water-environment/environmentally-friendly-city-initiatives/transformto/

City of Toronto, 2019b. ResilientTO. Available at: https://www.toronto.ca/services-payments/water-environment/environmentally-friendly-city-initiatives/resilientto/

City of Vancouver, 2009. Passive Design Toolkit. Available at: https://vancouver.ca/files/cov/passive-design-large-buildings.pdf

De Rubeis, T., Nardi, I., Ambrosini, D. and Paoletti, D., 2018. Is a self-sufficient building energy efficient? Lesson learned from a case study in Mediterranean climate. *Applied Energy*, 218, pp. 131–145. doi:10.1016/j.apenergy.2018.02.166.

Dezeen, 2019. 10 off-grid homes for a self-sufficient lifestyle. *Dezeen*. February 16. Viewed on February 19, 2019. Available at: https://www.dezeen.com/2019/02/16/off-grid-self-sufficient-homes-sustainable-architecture/

De Jong, M., Joss, S., Schraven, D., Zhan, C. and Weijnen, M., 2015. Sustainable–smart–resilient–low carbon–eco–knowledge cities; making sense of a multitude of concepts promoting sustainable urbanization. *Journal of Cleaner production*, 109, pp. 25–38. doi:10.1016/j.jclepro.2015.02.004.

Dunne, D. 2018. Geoengineering carries "large risks" for the natural world, studies show. *Carbon Brief*, January 22 Viewed on April 16, 2019. Available at: https://www.carbon-brief.org/geoengineering-carries-large-risks-for-natural-world-studies-show

EPA, 2017. Sustainable Management of Construction and Demolition Materials. United States Environmental Protection Agency Official Website, Viewed on February 14, 2019. Available at: https://www.epa.gov/smm/sustainable-management-construction-and-demolition-materials

Foster, N., 2015. Designing urban infrastructure: Investing for now or tomorrow, *Urban Age*, published on December 5, viewed on February 24, 2019. Available at: https://youtu.be/-GSfo4dTjk0

Fullychargedshow, 2017. *Sustainable City, Fully Charged*, online video, 17 March, Viewed February 2017, https://youtu.be/WCKz8ykyI2E

Galli, A., Iha, K., Halle, M., El Bilali, H., Grunewald, N., Eaton, D., Capone, R., Debs, P. and Bottalico, F., 2017. Mediterranean countries' food consumption and sourcing patterns: An ecological footprint viewpoint. *Science of the Total Environment*, 578, pp. 383–391.

Geddes, P. 1915. *Cities in Evolution: An Introduction to the Town Planning Movement and to the Study of Civics*. London, Williams & Norgate.

GEF, 2018. Catalyzing Solutions for Sustainable Cities. Viewed February 6, 2019. Available at: https://www.thegef.org/publications/catalyzing-solutions-sustainable-cities

Gervásio, H. and da Silva, L.S., 2012. A probabilistic decision-making approach for the sustainable assessment of infrastructures. *Expert Systems with Applications*, 39(8), pp. 7121–7131. doi:10.1016/j.eswa.2012.01.032.

GPSC, 2018. Catalyzing Solutions for Sustainable Cities. Viewed on February 6, 2019, Available at: https://www.thegpsc.org/knowledge-products/urban-management/catalyzing-solutions-sustainable-cities

Heath, G.W., Brownson, R.C., Kruger, J., Miles, R., Powell, K.E., Ramsey, L.T. and Task Force on Community Preventive Services, 2006. The effectiveness of urban design and land use and transport policies and practices to increase physical activity: A systematic review. *Journal of Physical Activity and Health*, 3(s1), pp. S55–S76. doi:10.1123/jpah.3.s1.s55.

HSE, 2018. What the law says. *Health and Safety Executive*. Available at: http://www.hse.gov.uk/temperature/law.htm

Houston, D. and Zuñiga, M.E., 2019. Put a park on it: How freeway caps are reconnecting and greening divided cities. *Cities*, 85, pp. 98–109. https://doi.org/10.1016/j.cities.2018.08.007

Jackson, D., 2019. Digital Earth: The paradigm now shaping our world's data cities. *The Conversation*, January 1, Viewed January 12, 2019. Available at: https://theconversation.com/digital-earth-the-paradigm-now-shaping-our-worlds-data-cities-104938

Kearns, P., 2012. Learning cities as healthy green cities: Building sustainable opportunity cities. *Australian Journal of Adult Learning*, 52(2), pp. 368–391.

Kılkış, Ş., 2016. Sustainable development of energy, water and environment systems index for Southeast European cities. *Journal of Cleaner Production*, 130, pp. 222–234.

Kurle, P., 2017. Cost implications of self sufficient and energy efficient urban residential building certification process. *International Journal of Engineering Science and Computing*, 7(5), pp. 11275–11280.

Landry, C., 2012. *The Origins & Futures of the Creative City*. Gloucestershire, UK: Comedia.

Landry, C., 2016. *The Digitized City: Influence & Impact*. Gloucestershire, UK: Comedia.

Layzer, J., Corburn, J., Thompson, J. P., Zegras, C., and Najam A., 2013. Sustaining Cities: Environment, Economic Development, and Empowerment. MIT World Series: Changing Cities: Celebrating 75 Years of Planning Better Futures at MIT, Viewed on February 2019, Available at: http://videolectures.net/mitworld_vale_layzer_corburn_thompson_zegras_sc/

Lejano, R.P., 2019. Climate change and the relational city. *Cities*, 85, pp. 25–29. doi:10.1016/j.cities.2018.12.001.

LEED Canada, 2010. LEED Canada for New Construction and Major Renovations. Available at: https://www.cagbc.org/cagbcdocs/LEED_Canada_NC_CS_2009_Rating_System-En-Jun2010.pdf

Liu, L. and Jensen, M.B., 2018. Green infrastructure for sustainable urban water management: Practices of five forerunner cities. *Cities*, 74, pp. 126–133. doi:10.1016/j.cities.2017.11.013.

Malalgoda, C., Amaratunga, D. and Haigh, R., 2013. Creating a disaster resilient built environment in urban cities: The role of local governments in Sri Lanka. *International Journal of Disaster Resilience in the Built Environment*, 4(1), pp. 72–94. doi:10.1108/17595901311299017.

March, H. and Ribera-Fumaz, R., 2016. Smart contradictions: The politics of making Barcelona a Self-sufficient city. *European Urban and Regional Studies*, 23(4), pp. 816–830. doi:10.1177/0969776414554488.

McPhearson, T., Andersson, E., Elmqvist, T. and Frantzeskaki, N., 2015. Resilience of and through urban ecosystem services. *Ecosystem Services*, 12, pp. 152–156. doi:10.1016/j.ecoser.2014.07.012.

Minaei, N., 2013. *Implications of Energy Use and How to Reduce the Impact*, University Lecture, Royal Agricultural University, Cirencester, UK, December 17, 2013.

Minaei, N., 2017a. Place and Community Consciousness. In *Smart Urban Regeneration*. S. Huston, (Ed.), pp. 68–84. Routledge and Taylor & Francis Group. https://www.taylorfrancis.com/books/e/9781317388432/chapters/10.4324%2F9781315677521-5

Minaei, N., 2017b. Smart Cities, Passive Design Strategies and Green Buildings, Sustainable Smart Cities (class lecture, University of Windsor, Windsor, ON, March 22, 2017).

Moore, T., 2014. From Intelligent to smart cities, *Australian Planner*, 51(3), pp. 290–291. doi:10.1080/07293682.2013.810163

Natural Resources Canada, 2016. Energy Fact Book 2012–2016. Available at: https://www.nrcan.gc.ca/sites/www.nrcan.gc.ca/files/energy/files/pdf/EnergyFactBook2015-Eng_Web.pdf

Natural Resources Canada, 2017. Energy Fact Book 2016–2017. Available at: https://www.nrcan.gc.ca/sites/www.nrcan.gc.ca/files/energy/pdf/EnergyFactBook_2016_17_En.pdf

O'Dwyer, E., Pan, I., Acha, S. and Shah, N., 2019. Smart energy systems for sustainable smart cities: Current developments, trends and future directions. *Applied Energy*, 237, pp. 581–597. doi:10.1016/j.apenergy.2019.01.024.

Odum, H.T., 1988. Self organization, transformity, and information. *Science*, 242, pp. 1132–1139.

Oxford Dictionary, 2019. Definition of self-sufficient in English, https://en.oxforddictionaries.com/definition/self-sufficient

Parris, K.M., Amati, M., Bekessy, S.A., Dagenais, D., Fryd, O., Hahs, A.K., Hes, D., Imberger, S.J., Livesley, S.J., Marshall, A.J. and Rhodes, J.R., 2018. The seven lamps of planning for biodiversity in the city. *Cities*, 83, pp. 44–53. doi:10.1016/j.cities.2018.06.007.

Pitt, A. 2019. Rockefeller Foundation to end funding of 100 Resilient Cities. *Cities Today*. April 8. Viewed on April 8. https://cities-today.com/rockefeller-foundation-climate-change-funding/

Soflaei, F., Shokouhian, M. and Zhu, W., 2017. Socio-environmental sustainability in traditional courtyard houses of Iran and China. *Renewable and Sustainable Energy Reviews*, 69, pp. 1147–1169. doi:10.1016/j.rser.2016.09.130.

Tillie, N. and van der Heijden, R., 2016. Advancing urban ecosystem governance in Rotterdam: From experimenting and evidence gathering to new ways for integrated planning. *Environmental Science & Policy*, 62, pp. 139–144. doi:10.1016/j.envsci.2016.04.016.

Toronto Park and Tree Foundation, 2019. Tree for Me. Available at: https://www.treeforme.ca/plant

Yigitcanlar, T., Dur, F. and Dizdaroglu, D., 2015. Towards prosperous sustainable cities: A multiscalar urban sustainability assessment approach. *Habitat International*, 45, pp. 36–46. doi:10.1016/j.habitatint.2014.06.033.

Yigitcanlar, T., Kamruzzaman Md, Foth M, Sabatini J, da Costa E, and Ioppolo G. 2019. Can cities become smart without being sustainable? A systematic review of the literature. *Sustainable Cities and Society*, 45, pp. 348–365. doi:10.1016/j.scs.2018.11.033.

UN Environment, 2017. Coastal crisis: Mangroves at risk. Viewed on February 18, 2019. Available at: https://www.unenvironment.org/news-and-stories/story/coastal-crisis-mangroves-risk-0

UN Environment, 2019. The sky's the limit as architects design UN17 eco-village in Copenhagen. Viewed on February 5, 2019. Available at: https://www.unenvironment.org/news-and-stories/story/skys-limit-architects-design-un17-eco-village-copenhagen

UNHABITAT, 2016a. Resilience. *UNHABITAT, for a Better Urban Future*. Available at: https://unhabitat.org/resilience/

UNHABITAT, 2016b. World Cities Report. Urbanization and Development: Emerging Futures. Available at: http://wcr.unhabitat.org/

Wakeland, W., Cholette, S. and Venkat, K., 2012. Food transportation issues and reducing carbon footprint. In *Green Technologies in Food Production and Processing* (pp. 211–236). Springer, Boston, MA.

Webb, J., Hawkey, D. and Tingey, M., 2016. Governing cities for sustainable energy: The UK case. *Cities*, 54, pp. 28–35. doi:10.1016/j.cities.2015.10.014.

World Bank Sustainability Blog, n.d. Can transit-oriented development change travel behavior in cities? *SmartCitiesDive*. Available at: https://www.smartcitiesdive.com/ex/sustainablecitiescollective/can-transit-oriented-development-change-travel-behavior-citie/1193359/

WHO (World Health Organization), 2009. Zagreb Declaration on Healthy Cities. Available at: www.euro.who.int/__data/assets/pdf_file/0015/101076/E92343.pdf

10 The Fifth Element
Biodigital and Genetics

Alberto T. Estévez

CONTENTS

10.1 INTRODUCTION: THE FIFTH ELEMENT

In the context of this book, "air, water, food, and energy—the four life-supporting elements," it can quickly come to mind, which in one way or another has fluttered by human thought since the beginning of history, the fifth element.

For some generations, this will remind one of the fantastic film by Luc Besson, *The Fifth Element* (1997), a classic of science fiction, not without symbolism, to be also applied to our days. It takes place in the twenty-third century, and it deals with the survival of planet Earth against a great evil that is approaching. Such salvation can only be achieved if four stones that contain the essences of the four mythical elements (air, water, earth, and fire), are joined. And these must be activated with the fifth element, which, in the form of a humanoid (genetically reconstructed from only a piece of living cells), is able to combine the power of the other four into a "Divine Light."

But curiously, at the end, the humanoid no longer wants to save the world because she has seen the brutality of human violence and refuses to release the Divine Light: "What's the use of saving life when you see what you do with it?" she asked. Luckily, the protagonist makes her feel something for what she does, love, which will be discovered precisely as the authentic fifth element. (In the background, there is also a skeptical look at capitalist consumerism, at the obsession with technology, and at the same time the tension in front of it that must be solved: nothing far from our present).

Actually, human violence is nothing new because it could be said—symbolically— that it exists since there were more than three people in the world. So the new great

195

evil, genuinely of our times, is the planetary unsustainability. And the survival of planet Earth against this great evil that is approaching can only be achieved joining really global efficient solutions for air, water, food, and energy.

As the four elements proposed by the Greeks are air, water, earth, and fire, it can also be said that "the four life-supporting elements" are "air, water, food, and energy." Understanding food as earth: after all food is a kind of earth transformation. And understanding energy as fire: after all, fire is pure energy and, symbolically, can be said that fire is the primary original energy, the best symbol for energy. Furthermore, they are not separate elements; necessarily they have certain connections between them.

But the four physical elements must be activated in a right way with the fifth element, which at Luc Besson's movie comes under the form of a humanoid; in our case, only humans can do it and have the responsibility for do it because only humans are to blame of what is happening. The humanoid is genetically reconstructed from only a piece of living cells; and in our time, we are discovering day by day how big the potential of genetics is. Genetics needs to be managed with intelligence, which symbolically is a kind of "Divine Light," able—in the film—to combine the power of the other four elements; the four elements that are life-supporting, as described in this book. And love, as conclusion of *The Fifth Element*, that in mankind is also life-supporting: the fifth element as humanoid, as intelligence, as "Divine Light," as love … in the movie as in the reality.

Obviously, if we talk about a fifth element, it is because before there is a certain agreement on four elements. Although, due to scientific advances in our contemporary age, speaking about the four elements has come to have something of poetic symbolism. However, an interesting symbolism to help the analysis of what happens today and, thus, conveys a synthesis that is useful to us, to understand our existence, and to achieve solutions to our current problems.

It is a fact that the considerations around the four elements, and something that can be ascribed to a fifth element, has gone through all historical eras and human traditions: From the oldest writings of the Bible ("the Word," "a mighty wind," "the breath of life") to the most ridiculous esotericism we see today on the Internet. Greeks, medieval Europeans, Hindus, Buddhists, Chinese, Japanese: *aether*, *quinta essentia* (from Latin, literally translated, fifth essence), *akasha*, void, spirit, word, wind, blow, breath, light, life, live, love. … Call it whatever you want.

Without writing a history of this topic, see one of the main alchemy treatises of all times for its complexity and for its erudite composition, the *Book of the Secrets of Nature or of the Quintessence of Ramon Llull* (1989), written in the thirteenth century. A title that curiously also recalls the contemporary words of Antoni Gaudí, when he was "searching in nature its mysterious image and turn it into architecture, forging the shape of the idea, discovering the laws of the universe in all its secrets." Something inspiring for our own research, in the understanding that nature has all the answers that our world needs for its sustainability.

It is clear that there must be something, when the classical four elements and the fifth element even appear frequently used in pop culture, music, literature, comic books, video games, movies, etc. It is something that fits well in the brains and thinking of all the people but also considering that, of course, the idea of the fifth element has actually been changing from generation to generation. So, it is time to say what would then be the idea of the fifth element for the present, that of the third decade of the twenty-first century.

10.2   BIODIGITAL AND GENETICS

We know that the answer is in nature and nature is the answer. The more the science advances, the more we know of what we call nature and the more we understand that nature is the answer. But, "if nature is the answer, what was the question?" (Wagensberg, 2008). We are exploring and interrogating "the question" through interdisciplinary endeavors involving fields such as material science, biology, genetics, art, architecture, civil engineering, design, computer graphics, and human-computer interaction. We are exploring the frontiers of knowledge. One main interdisciplinary cross point for really arriving at these frontiers is where genetics meets biology and the digital, applied to architecture and design in our case (and also applied to art or civil engineering, etc.). This is the cross point at which we find ourselves, the cross point that this chapter is about.

Things that we do not attach importance to or are scattered and disconnected sometimes end up converging in a total picture where all those previous aspects can be appreciated together. Or, perhaps, it is just the previous existence of these scattered aspects that allows us to reach the point where we are today. So, perhaps because of these previous latent interests, which should be explained in a different (autobiographical) framework, the Genetic Architectures Research Group and Office and the Biodigital Architecture Master Program was founded at the ESARQ, the School of Architecture of Universitat Internacional de Catalunya (UIC) Barcelona. This is the scene where we work to create architecture and design with geneticists focused on architectural objectives and architects researching the fusion of biological and digital techniques.

Perhaps it is not a coincidence that in this same city of Barcelona, that of Gaudí, where Salvador Dalí (1956) prophesied that "the architecture of the future will be soft and hairy (furry)," the year 2000 saw the real application of genetics to architecture with the creation of the first real genetic architecture laboratory on the planet, with the creation of the first digital production workshop in a Spanish school, and with the creation of the first official postgraduate, master's, and doctoral program on these genetic-biodigital issues.

Certainly with this the aforementioned Dalí prophecy is fulfilled. More when he realized that for it was simply a matter of having the necessary technique, the one from genetics, which took half a century more to arrive: "in 1925 I met Mr. Le Corbusier. … he asked my opinion regarding the future of architecture. I replied that I saw her 'soft and hairy (furry)'. I have not changed my mind and I hope that the technique reaches me one day, because I'm still a bit ahead of it." (Dalí, 2003). Although starting from a surrealist *boutade*, he was not mistaken, corroborated today in the shape of "biodigital," genetic, and "tidied"—enabled—by means of computing. Both for architecture and design, as for any other field.

It all started with a word. And after—let's say—an infinite succession of words, which started emerging as calls from the darkness of nowhere, this word—*biodigital*—emerges.

Truthfully, more and more—no doubt about the way forward—we can see how today biodigital is the future: the houses, the cities, and the landscapes will be 50% biological and 50% digital and the fusion of them both. It is certainly the great potential of new biological and digital techniques that can lead to the sustainable and social efficiency the planet needs, so the human being can have a future.

And, biodigital has in itself its own set of words. At the same time each one of them is related—at least, neurologically—to many others. It is in this way that around the term *biodigital* we find others such as biology, life, computing, nature, cybernetics, genetics, mathematics, DNA, algorithms, emergence, morphogenesis, artificial intelligence, surrealism, digital organicism, genetic architecture, robotics, biomanufacturing, digital production, bio-learning, programming, scripting, parametric, among others.

This cloud of words, in constant change depending on how one aspect or another is intensified, like if it were a pointillist mosaic, ends up illustrating what the term *biodigital* really is and what it can be (Figure 10.1). Biology and digital: if we also understand working with DNA as if it were a natural software, and with the software as if it were an artificial DNA, they are the new materials of the future, as concrete and steel were for modernism.

10.2.1 A Bit of (Genetic) History

It might now make a little sense to give a little history on the subject: between December 1999 and January 2000 a snippet of information went viral with the media publishing increasingly more information on it and constantly "infecting" each other. In this case, it was about genetics: press, radio, and television were quickly inundated by news and reports on this subject.

Then, watching how genetics offered such a huge field in the world of health and nutrition, I wondered about the application of genetics to architecture and design. For that reason, we created (already more than 15 years ago) the aforementioned research group and master's degree program and also connected it to a doctoral degree program.

Without knowing it yet, at the same time (2000) Eduardo Kac was working on Alba, with the green fluorescent protein (*GFP*) gene responsible for bioluminescence, becoming the most famous bunny of the twenty-first century. As is it known, this gene encodes the synthesis of the GFP protein, which is widely used in genetics as a marker, an indicator that allows for easy verification of the genetic transformation success. And though it is now supplied to genetic laboratories without problems, the natural source of this gene was originally a jellyfish called *Aequorea victoria* from the Northeast Pacific, in which the *GFP* is responsible for glowing in the dark. After the transformation of the cells that have been required for each case, the gene synthesizes the protein, allowing the cells to emit a bright green color when exposed to blue or black light. However, the GFP is also present in hundreds of sea species, with green, orange, and red colors in sharks, eels, seahorses, fish, coral, etc. This discovery has recently given rise to *fluo diving*, night diving in fluorescent underwater marine life, as if one were floating in the *Avatar* movie. *Avatar* (2009), another classic of science fiction by James Cameron, not without symbolism, to be also applied

FIGURE 10.1 Example of biodigital architecture and design. Photos and images by Alberto T. Estévez—GenArqOffice. Top left, *multifunctional building and park*, Hard, designed from the design strategy of digital voronoi structures. Bottom, photos of biological voronoi structures taken with scanning electron microscope (SEM; pollen, 1600×, and rose petal, 800×). Top right, *park*, Cornellà del Llobregat, designed from the design strategy of structures of attractors and digital force fields: presented on photos of the biological linear structures of wood. Bottom, *telecommunications building*, Santiago de Chile, designed from the design strategy of digital fractal structures: right, photos of biological fractal structures of dandelion, in three different scales; the lower one taken with electron microscope (pollen, 200×).

to our days, shows a whole planet—Pandora—in the way that we were researching from 2003, when we started the research about bioluminescence.

One day in January 2003, talking with the geneticists in our group about the use of GFP in research, Dr. Miquel-Àngel Serra asked "what else can the GFP be used for other than being an indicator?" As an architect it was clear to me: "for illuminating architectural spaces!" At that moment we began research for getting trees to work as "lamps," illuminating streets, plants homes, vegetation illuminating the roadsides without electricity; hence, the creation of plants with natural light by genetic transformation for urban and domestic use had emerged.

So, in October 2005, thanks to our geneticist Dr. Agustí Fontarnau along with Dr. Leandro Peña, we successfully obtained the first seven lemon trees with

FIGURE 10.2 *Genetic Barcelona Project*: The magic light of the green fluorescent protein (GFP) lemon trees. Center, image of a possible world. *Right*, real comparison between a lemon tree leaf with GFP and another without GFP from the same tree type. Top photo taken with conventional reflex camera, and bottom photo taken with special ultraviolet camera (author's images and photos).

luminescent leaves (Figure 10.2), provided by GFP. These transformed lemon trees get their green fluorescent protein through the expression of the *GFP* gene. That gene was transferred to the lemon tree cells into a transforming in vitro culture experiment using a DNA vector containing *GFP* genes (the DNA containing the *GFP* gene was not spliced; it was inserted into the lemon tree genome and kept intact inside of it): some of the transformed cells regenerated a new plant with its cells expressing GFP, knowing that the glowing properties can be seen by microscope from the beginnings of its cellular transformation. In 2 months, the trees were approximately 30 cm high, so that we were able to directly see the bioluminescent properties with our own eyes, taking photos with blue light and a conventional reflex camera, or along with Dr. Josep Clotet, also from our university, taking photos with our special ultraviolet (UV) camera.

We started with the GFP because it is the most studied one; geneticists use it as a common cellular marker. The functionality of the objectives was clear: The trees in this project were made with the objective of being of architectonical and urban use; it was the first time in architectural history that geneticists had worked for an architect; in 2005 we also presented this research under the name of the *Genetic Barcelona Project* to the mayor of Barcelona.

The durability results were good: today, more than 13 years later, the leaves have the same luminescence, and the initial little lemon trees continue to grow depending on soil availability. They can be also multiplied by planting their branches, becoming nonmanufactured "lamps," for free! But from the beginning, the lightning efficiency was poor and needed special light input to achieve enough brightness.

So, through a second phase of this project, to make more efficient and useful bioluminescent vegetation, we arrived at "Biolamps" (Figure 10.3): In 2007 we started to research bacterial bioluminescence for urban and domestic use. We also were involved with the research of how to achieve bioluminescent plants with a bacterial genes group that are the responsible for bioluminescence at the same time.

FIGURE 10.3   "Biolamps": the first systematically fully illuminated apartment with living light (human eye view: photos by the author, taken with a conventional reflex camera).

In 2008 we began to create Biolamps, a kind of "battery" with bioluminescent bacteria that are originally found in abyssal fish. With them, we created the first systematically fully illuminated living light apartment without electricity. For the first time in architectural history—without an electrical installation—a whole home was illuminated using bioluminescence. And as in Pandora, it is that the night lighting that we really need just to "mark" the environment and who circulates through it. It is needless the enormous waste of light energy that we spend today illuminating things and whole cities. The fact that we dazzle ourselves uncomfortably when seeing a focus already tells us how unnatural our system is. It is enough that things, walkways, remain "marked" with light.

The digital design and manufacturing of the Biodigital Lamps Series and their use as Biolamps had also begun (Figure 10.4). These lamps are based on an analysis of radiolarian structures and pollen. This analysis was applied to the digital development of architecture and design, by first using a scanning electron microscope (SEM; we have used it since 2008). This continues along the lines with the idea of "bio-learning," which offers the benefits of the structural, formal, and processual efficiency that we can learn from nature. Using CAD-CAM technology, and

FIGURE 10.4   Alberto T. Estévez—Genetic Architectures Office, *Biodigital Lamps Series*, being used as "Biolamps" digitally three-dimensional printed.

once we consider that the drawings have reached the desired result, will we proceed to its digital manufacturing, directly on a scale of 1:1. In this case, we can take advantage of different parts or levels where this technology allows for interscalarity. This allows us to easily change the scale of the jewelry and lamps to that of the pavilion. Research starts by choosing a system and a structural, architectonical, and design idea using geometry to draw it. Finally, to manufacture this, research with digital machines needs to be carried out with the confidence that "what can be drawn can be built" (Figure 10.5).

Paradoxically, the second phase of this bioluminescence research was effective for lighting, not as the GFP lemon trees, but too problematic in terms of durability; every 10 days the "bio-batteries" needed to be changed. The other option was fabricating a lamp that could guarantee the required air tightness, oxygen, and food for the bacteria. However, it was determined that the lamp was too complicated to manufacture compared with a simple bioluminescent plant or tree; our research includes a "democratical" goal, to achieve something for the whole society, and not only for those who have high purchasing power.

We are now in the third phase, trying to introduce the genes responsible for bioluminescence in ornamental plants. First, we obtained two stable lines using bioballistics with the plasmid pLDLux integrated into the genome of the *Nicotiana tabacum W38* chloroplast, and we can assert that the expression of the bacterial operon luxCDABE is correct and stable. We have also done the same using bioballistics with different species of ornamental flowers as *Begonia semperflorens, Codariocalyx motorious, Mathiola incana, Dianthus caryophyllus* (Figure 10.6). However, because of the low bioluminescence provided by the pLDLux vector (probably due to a lack of *LuxG* gene whose mission is to participate in the turnover of flavin mononucleotide (FMN), our efforts are now being focused on finalizing a vector of chloroplast transformation possessing *LuxCDABEG* genes—a work in progress!

After its presentation in congresses and publications in 2005, 2006, 2007, and forth,[1] we can say that the diffusion of this research has been a success. In 2010, the US "Bioglow" company took our idea for producing plants that can illuminate

FIGURE 10.5 Some images of the drawing process and digitally manufactured pavilion (right, author's photo of the *Biodigital Barcelona Pavilion* and previous drawings with the collaboration of Daniel Wunsch).

FIGURE 10.6 Author's photos of the current research with different species of ornamental flowers already genetically transformed but with too low bioluminescence.

human spaces. Soon thereafter, in 2012, the "Glowing Plant Project" began to search for the same ones (but not without controversy). "Bioglow," led by a geneticist, has also seen this as a potentially powerful niche. The second one, the "Glowing Plant Project," led by a businessman, also saw this as a great business opportunity. Since then there have even been different cases when the mass media has occasionally come out with the "amazing" news about the "novel" idea of illuminating trees illustrated using Photoshop.[2]

10.2.2 Toward the Frontiers of Knowledge

We can see the enormous potential nature offers us to assure a better future for our planet. This is the path toward the frontiers of knowledge. After all these years of work, a big difference remains between what we can imagine and what we can achieve because everything depends on getting money for research.

For example, we have already opened three ways of bioluminescence to drastically reduce energy consumption for night lighting and the pollution it produces. We are now at the threshold of a fourth possibility, which can be much more effective using bioluminescent fungi (we are preparing the identification of the responsible genes for bioluminescence of *Mycena*, *Gerronema*, and *Armillaria*). A fifth possibility is researching bioluminescent yeast, which is more experimental and perhaps more spectacular.

This is, thereby, the beginning of a revolutionary change in the cultural understanding of light, city, and architecture. This is also applicable to heat and habitat. What is at the end of the road? The satisfaction of meeting three of the most basic humans needs solved in the most natural and sustainable way: natural light, heat, and habitat and living without consuming energy and producing pollution, fueled by the ancestral power offered by nature. Trees and plants can offer biolight and bioheat through the most natural way for streets and homes; there are even vegetable genes responsible for providing warmth. Imagine living biohouses like trees or mushrooms with inhabitable conditions that can be purchased in malls; seeds that can be planted in the ground and grow alone; this all opens up an infinite unexplored horizon.

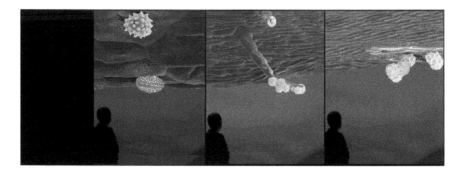

FIGURE 10.7 Author's images of biolamps and bioceilings based on research with scanning electron microscope (SEM).

How can we visualize future cities and future houses? As "soft and furry (hairy)" architecture, living cities and houses (Figure 10.7). As has been said before, the city of the future will be 50% biological technology and 50% digital technology (100% biodigital). Living houses that grow alone, trees that give light at night, and plants that provide warmth in the winter: a city that is more like a forest than a landscape shipping containers on the port. After all, where do we prefer to live, in boxes or in trees? Our cities are destroying nature wherever they grow. We need to assure that every human footprint becomes a creator of life. We need to change our reality with life!

10.2.3 (Gen)ethics

In our triplet of research-teaching-profession we work from (gen)ethics, yes, from responsibility, too, without forgetting that "humanity has the responsibility to have a future" (*It's not enough! Manifesto*, New York, Autumn 2010), for which we must already urge a nonconservationist vision of nature: the human beings, in the first instance, need to overcome such vision to survive. It is plasticity that the most characteristic feature of nature. To the point that it would be against nature to want to "freeze" life in each of its always changing appearances already known. Nature is not an exotic collection of diverse species. It is more the fluidity with which it always presents itself under diverse infinite facets. And the human being is its most powerful vector. "Make it flow!" (*Bioplasticity Manifesto*, Barcelona, Spring 2010).

Genetic research for architecture also requires precautions—by avoiding accidents and contamination—like in conventional medical research or in simple heart surgery. Science requires responsibility, and we are establishing strict procedures for testing in hermetic environments, breeding plants without pollen, or by acting in chloroplast to avoid pollination problems. Our team includes philosophers dealing with bioethical matters, like Dr. Josep Corcó and Dr. Xavier Escribano. We hyphenate the word "gen-ethics" meaning "ethics" in our research when the need for planetary sustainability justifies our work.

Nonetheless, from an objective point of view, there is no ethical difference between acting on "the surface of things" and acting at the intramolecular level. Once we accept the organic and fluid configuration of nature, there is, ethically speaking, not much difference between the production of a Japanese bonsai and a fluorescent rabbit. Bonsais are socially accepted even though they are the result of "tormented" living matter, and a fluorescent rabbit is not less happy than a black or white one.

What is more, the most extreme action would be eating a living being because we simply do not kill it but instead we make it disappear into our own cells. However, nobody is put in jail for eating a chicken sandwich. Because this applies to even the most extreme action—eating—it automatically follows that any other less drastic action is permissible, excluding ill treatment.

But of course, if we work with genetic material we must accept our responsibility as illustrated by the "domino effect" that takes place in time and space and that has been explained using the example of a butterfly. In spatial terms, there is the "butterfly effect," where the beating of a butterfly's wings in China is said to be able to trigger a storm in United States. In terms of time, we can refer the dramatic book *A Sound of Thunder* (Bradbury, 1953) in which a prehistoric butterfly is accidentally killed by a traveler from the future, thereby changing life millions of years later.

It is not only our actions on genetic material but, all of our actions, that have a corresponding domino effect millions of years later. At least everything is part of nature, but with genetics "a new and vast territory is removed from the realm of randomness and enters into the realm of morality. We are captives of our own competence, of our own capabilities, by which we recreate what we only wanted to represent, or we transgress the natural order that we only pretended to repair" (Rubert De Ventós, 2015).

10.2.4 KAC'S AFFAIR

Furthermore, the way Eduardo Kac explains his work should be approached in terms of ethics. He defined *transgenic art* as "a new art form based on the use of genetic engineering techniques to transfer synthetic genes to an organism or to transfer natural genetic material from one species into another, to create unique living beings" (Kac, 1998). The only wrong and confusing aspect of this definition is the word *unique*, like when he claims to be some kind of God-Creator and oversteps the definitions that humans have agreed on. This transgression does not do science any favors. The account that he likes to offer in public (like the following excerpt from an interview) does more harm than good:

> – it took—seven years!—of work on the Edunia petunia before I managed to introduce my own DNA into it… . I put my DNA into its 'veins', and now it is producing my human proteins. The green phosphorescent rabbit and the 'plantimal' aren't nature… . I created them! … With Alba (2000) and the plantimal Edunia (2003), I also relieve God of his status as a creator-myth and turn him into a lab worker, a technician working in a transgenic workshop.
> – You don't seem very humble.
> – I don't copy reality: I create it (Amiguet and Kac, 2012).

It has a negative effect that somebody with a strong presence in the media speaks without any scientific accuracy, demonstrating terminological and ideological confusion. In addition, the necessary clarifications and criticism are not arguments taken from the author's subjective point of view because they have a background that is substantiated by the previously mentioned scientists, geneticists, and philosophers in our research group:

> It is not true that he spent 7 years on this project; it simply took 7 years to happen.
>
> It is not true that he inserted his DNA into the plant; it was more like having a "microbrick" inserted into an enormous set of many thousands of microbricks. In any case, this microbrick is identical to the ones that we all have, and it is not in any sense specifically or uniquely "his."
>
> It is not true that the resulting plant produces "his" human proteins. Rather, it produces human proteins that are chemically identical to those of any human.
>
> It is not true that by inserting a gene taken from an animal into a plant, it becomes a "plantimal." Just like a virus can mutate our cell's DNA and cause a tumor, this does not make us a "humanirus."
>
> It is not true that the rabbit and the plant in question "are not nature."
>
> It is not true that he created this rabbit and this plant.
>
> It is not true that he relieves God of his status as a creator-myth and turns him into a lab worker, a technician in a transgenic workshop because the common definition of "God"—as human agreement—includes he who "creates from nothingness." Genetic manipulation simply involves repositioning existing microbricks.
>
> It is not true that he "creates reality" because the gene that he integrates into an enormous preexisting genetic structure already existed before. Therefore, he does not create a single gene; he simply changes its position.

Basically, by inserting a gene from another being into the rabbit and the plant, they did not cease to be "natural"; they did not cease to be nature. This gene "repositioning" has been carried out anonymously by the pharmaceutical and agricultural food industries long before Kac's projects, with more complexity and implications and on a large scale.

10.2.5 DECONSTRUCTING NATURE

However, this emergent character of life is what humanity has to take advantage of, and this is why we are interested in investigating how genetics can be applied to architecture. The idea is to take advantage of nature's capacity for self-organization, growth, and reproducibility "for free." Therefore, we look for plants that emit light or heat and will help find the energy-saving mechanisms that our world needs and that will be usable as construction materials and even as entire habitats. We can begin to imagine, in a not so distant reality, "streetlights," "heaters," and even entire houses that grow on their own.

Given that this research also focuses on the use of genetics, we can also consider possible architectural uses at the level at which undefined cellular masses emerge and self-organize, as the first structural step: research on the genetic level where masses of cells organize themselves into primigenic structures to find structures and to learn typologies that could be of interest for architecture, which will be illustrated as an alternative landscape of the future. We are studying this with a SEM, which

FIGURE 10.8 *Living city*: the enigmatic evocative power of scanning electron microscope (SEM) images (author's images).

has an extremely high resolution, allowing to us see images magnified thousands of times. This opens up a little -known dimension of reality, which, depending on how the images are read or interpreted, can lead to a fascinating level of surreality (Figure 10.8). As a result of research carried out in this framework, it was possible to create strange and surprising new images: "altered" photographs of natural structures at their most Genesis-like and primitive level. Artistic works and architectural plans based on biotechnological work that have an enigmatic evocative power. Always justified for being behind the first laboratory's effort that began the real application of genetics to architecture, fighting for the sustainability of our entire planet and a better world.

To conclude, describing the future development of our work and providing a reasonable projection of the research into future applications, it can be said that the equilibrium of our planet—for our own survival—needs several things: accurate and precise reset of our behavior, education, our basic habits, food, our ability to manage waste, and consumer goods. For example, we have to get used to the idea that we do not need so much light at night (our eyes have a wide range capacity that we do not use) like we put on a sweater if it is cold.

Consumption of energy must be radically reduced: A middle-sized European city of only 100 km² spends 10 million Euros annually just on the maintenance of its street lights (new lamps, repairs, repainting) in addition to electricity consumption. If multiplied by all the cities on the five continents, the figure is absolutely astronomical. Therefore, bioluminescence will substitute artificial lighting at some levels; there is no doubt about that. Nature is always teaching us, in this case with in many bioluminescent ways, from bacteria and plankton to algae, fungi, insects. It would be like suicide for the next generations if we did not learn about it.

And this is only a fraction of the possible scope of the application of genetics for architecture and design. When the architect stops needing conventional construction industries and starts working with geneticists, who are the bricklayers of the future, we may design evolving science, architecture, and design collaborations in which genetics becomes integral to architectural research and production with infinite possibilities.

The research into the architectural application of cutting-edge biological and digital techniques (with the benefits that come from the inclusion of genetics: efficiency, economy, renewable use, self-replication) is crucial, relevant, and urgent before it becomes too late for our planet, which has reached the limits of its sustainability. "We have, because human, an inalienable prerogative of responsibility which we cannot devolve" (Sherrington, 1940).

10.2.6 LEARNING FROM NATURE, ALSO FOR BEAUTY

Everything that has been explained so far comes from bio-learning. And bio-learning—of course—is contained in "learning from nature." Although the preposition used provides nuances: learning "from" nature also means learning "with" and learning "in" nature, even "knowing nature" itself, without any preposition. Each expression has its significant nuances, which should not be discriminatory but enriching when considering them as an interactive interrelated whole. This understanding leads to define genetic architecture as something that is no longer about building "in" nature, but building "with" nature, and even building nature itself, equally without any preposition.

On the other hand, it has been confirmed that nature is an eternal mirror for human aesthetics, as well as for its aspirations. Generation after generation, nature never becomes obsolete and it never tires. It has always been, is, and will be as perennial as an open book, unique and indivisible. Nature is an inexhaustible source of inspiration, imitation, or learning. Biodigital architecture and genetics, defined as directly involved in its incardination "with" nature, thus ensures aesthetic "durability." It could even be said that it is a guarantee of "classicity" and adapts to the times. Even more so when new techniques open up new fields that are still unexplored. We are living a great epic and heroic age. It is an age of opportunities where the brave and daring will launch themselves into the unexplored and become the pioneers of the biodigital and genetics age.

Thus, the closer the processes of architectural creation are to nature, the less obsolete and more "eternal" the result will be. It is necessary to listen to the language of nature and reply to it coherently if, in the end, nature and the entire universe are written in mathematical language, as Galileo Galilei suspected. We are talking about languages that are always valid and reduce the arbitrariness of our decisions when harmonizing them. Thanks to digital tools, this provides us with control, efficiency, and a harmonious accuracy that enables us to exclude arbitrariness as much as possible.

Everything can indeed be solved by learning (in depth) from nature. At all its levels, from the most "internal" and intramolecular one, accessible today thanks to genetics, to the most "external" and superficial one, which has also been imitated by human beings from the moment they came into existence. It is not a coincidence, for instance, that human beings are attracted to the sight of the air (clouds, smoke), water (the sea, waves), earth (rocks under the action of water and wind, geological crystals under the action of physical and chemical processes), and fire. It furthermore coincides with the four roots of Empedocles, the four primitive elements, which confirm how their changing forms permanently remain configured by actions or laws that affect the whole as well as every part alike. Architecture and design, which also follow similar laws, equally evoke a similar attraction: something invisible to the human eye that "from the inside floods" each cell, its entire appearance and even its most remote corners (continuity). It resounds in all its parts, configures the whole (Alberti's *concinnitas*) and inevitably controls its constant evolution (emerging system).

Why do we like looking at clouds (air), waves (water), cliffs (earth), and bonfires (fire)? We never tire of them because they calm us and attract us, and we all agree we perceive beauty, "objective beauty," in them. Furthermore, as they move, our interest becomes addictive. Their shapes do not bore us and because of their complexity and because they change (without us moving), they even surprise us: when each and every part responds to the whole because of objective laws, physical and chemical determinants, and genetic ones in the case of living beings that need to carry out specific functions; when "each part is reflected in the whole and the whole is reflected in the parts" (the most classic definition of beauty), an organic, organized continuous, coherent, united connection exists; and when each and every one of these words turns into a value for architecture and design, always moved, created by common external physical-chemical forces or internal ones driven by DNA.

When the determinants are almost purely and exclusively genetic, or at least still mostly genetic, when the consequences of a specific diet, habits, climatology, a specific and distinguishing genetic inheritance, or whichever other random external determinant are still not completely reflected, it is then when the emerging character of life driven by DNA "clearly shows" its own force more; it is then when unanimous, spontaneous, and popular qualifying adjectives such as "cute," "lovely," and "sweet" are on everyone's lips, which is something common when ones sees a puppy or a baby.

All this supports the "objective beauty" Gaudí talked about, when something has certain characteristics that make the definitions of beauty comply and that, in addition, coincide in qualifying it as such. However, in the time of Gaudí, genetics did not exist, and he, therefore, did not know about the consequences of the "natural computer," which is DNA. And, of course, he did not have any digital computers that could organize a complex and united whole, and, at the same time, measure it with absolute accuracy and control it. This is why he had to invent his own nondigital computers: catenary ropes hanging freely in space, which, thanks to the strategic position of little sachets filled with lead, could simulate to scale the real loads the building would have to support, ordering its lines "automatically," "parametrically"—lines the author did not directly and with pinpoint accuracy decide on but rather the "computer" supervised by him to configure an objective, harmonious, mathematical beauty.

"Objective beauty" thus turns into "necessary beauty" when it becomes a human need and a duty of architects and designers toward humanity, willing to create architecture and design in an equally complex way, that cannot be used up in the blink of an eye, nor be understood in a second, where every point of view is different (as we are the ones that move) and, therefore, awakens interest and responds to a coherent whole at the same time. It is nature that shows us the way to create and develop it.

10.3 CONCLUSION: INTEGRAL ECOLOGY

The four life-supporting elements are exactly this, supports of the fifth element, life! Understood at the same time in a literal sense and in a metaphorical sense. Paradoxically the four elements need the fifth element to be activated as life supports (the fifth element, life, intelligence, knowledge, ... call it whatever you want). This has a poetic point of view: We will solve our planetary problems with life, but it has also the exactly description of the real and scientific path to follow, through biodigital and genetics, life for saving life.

And, because everything is interrelated, and today's problems call for a vision capable of taking into account every aspect of the global crisis, we need to urgently consider the idea of integral ecology, one which simultaneously consider as an indivisible whole its environmental, economic, and social dimensions.

Ecology studies the relationship between living organisms and the environment in which they develop. This necessarily entails reflection and debate about the conditions required for the life and survival of society and the honesty needed to question certain models of development, production, and consumption.

Just as the different aspects of the planet—physical, chemical, and biological—are interrelated, so too living species are part of a network that we need to explore and understand, and genetics is the most advanced tool. Even a good part of our genetic code is shared by many living beings. It follows that the fragmentation of knowledge and the isolation of bits of information can actually become a form of ignorance, unless they are integrated into a broader vision of reality.

When we speak of the "environment," what we really mean is a relationship existing between nature and the society that lives in it. Nature cannot be regarded as something separate from ourselves or as a mere setting in which we live. We are part of nature, included in it and, thus, in constant interaction with it. Recognizing the reasons why a given area is polluted requires a study of the workings of society, its economy, and its behavioral patterns. Given the scale of change, it is no longer possible to find a specific, discrete answer for each part of the problem. It is essential to seek comprehensive solutions that consider the interactions within natural systems themselves and with social systems. We are faced with one complex crisis that is both environmental and social. Strategies for a solution demand an integrated approach to protecting nature and at the same time combating poverty.

We, as architects and designers, have the personal responsibility to achieve a vision of architecture that helps to develop sustainable and safe societies. And this goal, in our present reality, is not only relevant but also urgent. Hence, we have the personal responsibility to create and engage ideas of environmentally responsible architecture, that means at the same time socially responsible architecture, and the biodigital integration, biology and digital, is the most advanced tool for architects and designers. Biodigital, as a tool, but before, as an approach to architecture and design, itself as an understanding of architecture and design. While we are at this work, little by little, almost as if by magic, the understanding about what our times demand will grow. In an integral two-sided reality, on one side, seeing that architecture and design can improve the world by improving the lives of the least fortunate,

FIGURE 10.9 Alberto T. Estévez—GenArqOffice, *3D Printed Sahara House Project*, 2017. Integral ecology example, digital design and digital manufacturing, on site, using the desert sand itself, abstracting in the conception of the project elements of the vernacular architecture typologies, with building climate control by means of passive systems.

and on the other side, learning from nature's laws (bio-learning) and finding computation as the most powerful tool for really solve problems (Figure 10.9).

Yes, the path to follow is clear, in a world where everything is already connected, with an intimate relationship between the poor and the fragility of the planet: when it is necessary to develop critical thinking toward the new paradigm and the forms of power resulting from technology; when it is necessary to search for new ways of understanding economy and progress, the value of each person, and the human sense of ecology; and when there is an urgent need of sincere and honest debates, there existing a great responsibility in terms of politics both at an international and a local scale. Ultimately, when it is convenient to have a new lifestyle. This is all what the *Zeitgeist* calls us to do.

At least this world is our common home that sustains us, and we must guarantee its own sustainability; that it is ours, too. Everything that discriminates the integral ecology will not prevent the ruin of our entire world. There will only be a real future if there is one for all. Only working for an integral ecology, with, of course, proportional generosity and sacrifices on the part of each and everyone, we will find together a real salvation of our beautiful Blue Planet.

1. For example, see ESTÉVEZ, Alberto T. "Genetic Barcelona Project." *Metalocus*, n. 17. Madrid, Fall 2005, pp. 162–165; ESTÉVEZ, Alberto T. "Genetic Barcelona Project: Cultural and lighting implications." *Urban Nightscape 2006*. Athens: International Commission on Illumination, 2006, pp. 86–88; DOLLENS, Dennis and ESTÉVEZ, Alberto T. "The genetic creation of bioluminescent plants for urban and domestic use." *Leonardo*, n. 40 (1), February 2007, pp. 18 and 46.
2. For example, see Paul Rincon, *BBC News*, 24th January 2013, about a team of undergraduates at Cambridge University; Alexandra Daisy Ginsberg, *Restorative Design*, 9 May 2013, about the Massachusetts Institute of Technology; Katherine Brooks, *Huffington Post*, 30 March 2014, about a Dutch artist/designer/architect; etc.

NOTES

1 For example, see ESTÉVEZ, Alberto T. "Genetic Barcelona Project." *Metalocus*, n. 17. Madrid, Fall 2005, pp. 162–165; ESTÉVEZ, Alberto T. "Genetic Barcelona Project: Cultural and lighting implications." *Urban Nightscape 2006*. Athens: International Commission on Illumination, 2006, pp. 86–88; DOLLENS, Dennis and ESTÉVEZ, Alberto T. "The genetic creation of bioluminescent plants for urban and domestic use." *Leonardo*, n. 40 (1), February 2007, pp. 18 and 46.
2 For example, see Paul Rincon, *BBC News*, 24th January 2013, about a team of undergraduates at Cambridge University; Alexandra Daisy Ginsberg, *Restorative Design*, 9 May 2013, about the Massachusetts Institute of Technology; Katherine Brooks, *Huffington Post*, 30 March 2014, about a Dutch artist/designer/architect; etc.

REFERENCES

Amiguet, Lluís; Kac, Eduardo. "Eduardo Kac, creador del conejo fosforescente." *La Vanguardia*. Barcelona: Grupo Godó, February 4, 2012, p. 64 (obc).

Bradbury, Ray. "A sound of thunder." In Bradbury, Ray. *The Golden Apples of the Sun*. New York: Doubleday & Company, 1953.

Dalí, Salvador. *Conferencia-manifiesto Park Güell*. Barcelona: Archivo Cátedra Gaudí, 29 September, 1956. Quotation taken from the Salvador Dalí conference in the Park Güell of Antoni Gaudí, in Barcelona, 1956, whose manuscript is conserved in the collection Fundació Gala-Salvador Dalí, Figueras.

Dalí, Salvador. "Confesiones inconfesables." In *Obra completa, vol. II. Textos autobiográficos 2*. Barcelona: Destino, 2003 (1973), pp. 491 and 625.

Kac, Eduardo. "Transgenic art." *Leonardo Electronic Almanac*, 6 (11). Cambridge, MA: MIT Press, December 1998.

Llull, Ramon. *Libro de los secretos de la naturaleza o quinta esencia*. Madrid: Ediciones Doble R, 1989.

Rubert De Ventós, Xavier. 154: Transformar el destino en diseño. In Cortina, Albert; Serra, Miquel-Àngel (eds.). *¿Humanos o posthumanos?* Barcelona: Fragmenta, 2015.

Sherrington, Charles. *Man On His Nature*. Cambridge: Cambridge University Press, 1940.

Wagensberg, Jorge. *Si la naturaleza es la respuesta, ¿cuál era la pregunta?* Barcelona: Tusquets, 2008 (2002).

Index

Note: Page numbers in italic and bold refer to figures and tables, respectively.

For Product Safety Concerns and Information please contact our EU
representative GPSR@taylorandfrancis.com
Taylor & Francis Verlag GmbH, Kaufingerstraße 24, 80331 München, Germany

www.ingramcontent.com/pod-product-compliance
Ingram Content Group UK Ltd.
Pitfield, Milton Keynes, MK11 3LW, UK
UKHW021119180425
457613UK00005B/157